New Physics And The Mind

To order additional copies, please contact us.
BookSurge, LLC
www.booksurge.com
1-866-308-6235
orders@booksurge.com

New Physics And The Mind

Robert Paster

2006

New Physics And The Mind

Contents

Introduction

Ernest Solvay was a successful nineteenth- and early-twentieth-century Belgian industrial chemist and manufacturer who also had a lifelong interest in pure science. He deployed some of his riches sponsoring a periodic series of conferences, the Solvay Conferences, where the world's greatest scientists discussed and pushed forward developments in theoretical chemistry and physics.

The Fifth Solvay Conference, held in 1927, included exploration of the theory of quantum physics by the world's preeminent physicists. A debate developed—involving two of history's great physicists, Albert Einstein and Niels Bohr—about the meaning of quantum physics. Einstein did not accept Bohr's views that reality depends on observation and that physical processes can be random and indeterminate.[1]

Einstein was a realist. He believed that the physical world is real, and that what's real does not depend on who's observing it. And he could not accept the view of Bohr and Bohr's Copenhagen colleagues that observation, or consciousness, or the mind has any role in physics.

Einstein's view dominated mainstream twentieth-century physics.

But seventy years after the Fifth Solvay Conference, the organizers of an international symposium on the Quantum Future preface the symposium's lecture notes by observing that physicists have still not resolved questions about quantum physics' meaning, and that perhaps it is time to revisit this old debate: "'Standard' interpretations of quantum theory are not satisfactory and it is becoming increasingly apparent that an alternative physical theory may be needed."[2]

In this book, *New Physics and the Mind*, we will explore the progress of physics during the past century, both modern physics' great successes and its remaining mysteries. And we'll look at standard and alternative formulations for how these mysteries are to be solved.

Taking Quantum Physics Seriously

In his paper at the 1997 Quantum Future symposium, Berkeley physicist Henry Stapp notes that quantum theory created such a sharp break with established science that its developers stipulated that its most radical elements not be taken seriously as a description of how the world works. These radical elements are those that "contaminate" matter with mind. Stapp, among others, is of the view that "decontamination" has failed, and that it is time once again to take quantum theory seriously, to allow consciousness and the mind back into our deepest understandings of physics.[3]

There are numerous variations on Einstein's and Bohr's views on the nature of reality. Einstein's realism generally has its descendents in the great successes of reductionism, the complete understanding down to the smallest components of the physical world. These successes challenge the need for a changed route to understanding the physical world. But holists question reductionism, and suggest the need for understanding emergent phenomena, self-organization, entanglement; they claim evidence particularly in phenomena of physics observed in recent decades.

In *New Physics and the Mind,* we will explore a century of discoveries in modern physics, and how physics is understood today in light of these recent decades' discoveries.

You will need no background in science or mathematics for this book. In fact, one of the themes is to think like someone who has been presented at once with what is known today about physics, not someone who must slog through the challenges of gaining the academic physicist's base of knowledge piece by piece, as it has developed over the years, or as it is taught at universities.

What if you had the luxury of finding out by reading just a few hundred pages where physics stands after a hundred years of modern efforts? Would you see the same organizing themes as today's hard-working academic physicists see? Would you draw the same conclusions as to what's resolved and what remains a mystery?

Chances are, with the luxury of not having to study the works of thousands of physicists who have preceded you, you will not feel so constrained to see the future progress of physics so linearly extrapolated from the past. You will instead get to take a census of all that remains mysterious, and organize these mysteries in today's, not yesterday's, categories.

After twenty-six chapters, we'll speculate. In these twenty-six chapters, three of the book's four parts, we'll review developments in physics during the twentieth century, review what physicists have said about the role of the mind in physics, and find out what the last decade or two's developments have been in fields of new physics.

Then we ask: Do we view physics' core questions the same way as they are viewed in the canon of academia? Or is there a way to understand today's mysteries from a different point of view than that of majoritarian physics? And what is being said by serious physicists, perhaps viewed as rogue by many of their colleagues, who have developed these different views of new physics and the mind?

Physicists are working worldwide on the topics covered in this book. The breadth and volume of physics research underway today is astounding.

I estimate that I have reviewed a million pages of material to prepare this book—several thousand books, and several tens of thousands of technical articles published in scientific journals and electronic forums—in search of physicists' investigations of phenomena of new physics, and in search of physicists' understandings of the role in physics for the mind and consciousness. Obviously, a book of this length can be only the most summary-level synthesis of a million pages of great scientists' and thinkers' work.

Immerse yourself in the patterns of today's research—the quest to develop the correct understanding of quantum gravity, which will give us a theory of everything; the reductionist urge to look at the basic constituents of physics with finer and finer resolution. Are these the routes to our deepest understanding? Or do we need a step back and a holistic point of view?

Hopefully you'll persist long enough to participate in our own speculations, which you'll earn the right to do in this book's Part Four. Until then, I hope you enjoy this story line of a hundred years of modern physics, this review of thousands of researchers' and theorists' efforts and reported findings.

And this is a story told with a point of view, physics with attitude. An attitude that is the basis for Part Four's Hidden Physics Countdown of theories of new physics and the mind.

Notes on Reading *New Physics and the Mind*

You are reading about the most current developments among brilliant and knowledgeable masters of an esoteric branch of advanced science.

I have tried to limit the use of difficult scientific jargon, to keep the explanations clear and the English plain, to define terms as simply as possible before they are used, and to say only what needs to be said in service of the book's ultimate theme.

This theme is that two important strands of modern physics—that the mind is part of physics, and that a series of newly observed phenomena challenge physics' standard models—can be intertwined to create a radical new scientific understanding.

Along the way, you will find sections introduced with the phrase "What's Going On With..." You may feel free to skip these if the reading's too dense for you. These are detailed discussions, mostly of Part Three's new physics, which are there if you'd like more information, but which aren't essential to the principal flow of *New Physics and the Mind*.

The only test at the end of *New Physics and the Mind* is whether the story line was convincing enough for you to be intrigued about the possibility of a radical new physics of the mind.

Summary

A surprising strand of modern physics, since its inception early in the twentieth century, has involved links between physics and human consciousness. *New Physics and the Mind* explores these links, especially as they are informed by recent decades' discoveries of equally surprising new physics phenomena.

The book is divided into four parts: Entering the Twenty-first Century, Physics and the Mind, New Physics, and Speculations. Each part contains from five to fifteen chapters.

Here's the story line.

Part One, Entering the Twenty-first Century, brings all readers to a common footing. After all, some readers will be physicists. Others will have read a few or many popular-audience science books, perhaps even books entirely on point to the question at hand: what's going on at the forefront of modern physics? Still others will be intellectually curious but without much background in science.

So Part One begins with a plain-English summary of the theories of relativity and of quantum physics. Then it discusses how physicists feel about these theories, and what these theories suggest to physicists about what still needs to be resolved, what to study next. String theory, which mainstream physics views as the best shot at the twenty-first century's theory of everything, is presented next. But so are the dissenting views—views of physicists who claim that quantum physics forces us to a new sense of reality, and also views of the holists, who think that reductionism is taking us along the wrong path.

From the beginning of quantum physics in the early twentieth century, physicists have found, strangely, a role in physics for human consciousness and the mind. This role has in fact been too strange for mainstream physics, which has proceeded without the mind.

Yet from the beginning, some physicists have explored how physics and the mind interrelate. This is Part Two, Physics and the Mind.

Theories about physics and the mind simmered in the background for most of the twentieth century, but they heated up again especially with the 1989 publication by a prominent mathematician and physicist of *The Emperor's New Mind*, which placed consciousness and the mind at the center of modern physics' most important current questions.

At the same time, late in the twentieth century, a series of observations and theories arose, not related to the mind or consciousness, but purely on the agenda of mainstream particle physics and cosmology. Fifteen of these phenomena are discussed in Part Three, New Physics.

The term "new physics" has been used frequently in physics, in fact ever since the dawn of modern physics at the beginning of the twentieth century. But today new physics means phenomena that challenge the mid-twentieth-century's standard models of particle physics and cosmology.

Many of these new physics phenomena have become widely known even among nonscientists. Black holes, dark matter and dark energy, parallel universes, even extra dimensions have all achieved some prominence as items of general news.

Physicists throughout the world are seriously studying these phenomena as well as other phenomena that have not yet achieved the same level of general awareness. Many of these other phenomena bring quantum physics to the macroscopic level of everyday existence, and this is surprising, even disturbing. Quantum physics has been understood to operate only subatomically. Some physicists studying the phenomena of new physics think that radical revisions are required to our standard approaches to physics.

And Part Four, Speculations, is a countdown—a hidden physics countdown of ten radical theories of new physics. These theories, barely noticed by mainstream physicists, are the life work of a small number of radical theorists.

Counting down from Radical Theory #10 through Radical Theory #1, we find physicists who turn quantum physics on its head, who create new understandings of physics from elements of new physics, and who bring the mind and consciousness into central roles.

Radical Theory #1 incorporates every element of Part Three's new physics, and is also a theory of mind physics. This theory, far outside of standard physics, is presented in *New Physics and the Mind* as an alternative to today's mainstream approach to a twenty-first-century theory of everything.

That's the story line of *New Physics and the Mind*, taken as a whole. Each of the book's four parts will also begin with a summary of the part and its chapters.

Part One
Entering the Twenty-first Century

In this, the first of *New Physics and the Mind*'s four parts, we begin with a review of the progress of the science of physics during the past century. The central theme that has emerged has been to create a Theory of Everything, which unites relativity with quantum physics, and which creates a single understanding of all matter and forces.

The clear contender for the Theory of Everything throne is string theory, now generalized to M theory. But troublesome observations and questions, including questions about the nature of reality, consciousness, and the mind, keep other contenders alive.

Part One's arts & entertainment section: Ghosts.

***

Part One Summary

In this first of the book's four parts, we discuss the major trends of twentieth-century physics and identify contemporary physicists' consensus position as to physics' open questions and how we should go about addressing them.

Chapter 1. Twentieth-Century Physics. We discuss the three major strands of twentieth-century physics: special relativity, general relativity, and quantum physics.

Chapter 2. Report Card on Twentieth-Century Physics. Consensus views among physicists are presented as to the major accomplishments in physics through the end of the twentieth century.

Chapter 3. Entering the Twenty-first Century: String Theories. The most highly regarded proposal for physics' theory of everything is a theory in which the universe's most basic entities are vibrating strings. These strings vibrate in ten dimensions of space, not just our familiar three spatial dimensions. Their vibrations explain the elementary particles of matter and the forces of the universe, including the force of gravity, which has proven to be theoretical physics' most troublesome force.

Chapter 4. Ghosts. An entertaining digression about how we might experience extra dimensions of space, beyond our familiar three dimensions.

Chapter 5. Entering the Twenty-first Century: What Is Reality? Quantum physics incorporates a surprising dependence on observers, on intervention into physical systems by measurement, even on a role for consciousness. This is because the quantum model is that unobserved elementary phenomena proceed along multiple, parallel, superposed paths until they are observed or measured, at which point there is a collapse to a single path. This astounding phenomenon raises basic questions about the nature of reality.

Chapter 6. Reductionism, and Its Critics. The tendency of scientists to learn and to understand by breaking complex systems into component parts is considered a principal approach of the modern scientific method. There is some thought among scientists that there is a limit to how far we can take these reductionist methodologies, and that holistic approaches need to be introduced, especially if we are to develop a scientific understanding of consciousness and the mind.

1.
Twentieth-Century Physics

This introductory chapter is a whirlwind tour of a hundred years of physics. This will bring us up to the final years of the twentieth century and will allow us to assess: what are the consensus views of the world's physicists regarding the central questions of physics, and regarding what remains on the horizon for resolution during the twenty-first century?

***

The three great strands of twentieth-century physics were special relativity, general relativity, and quantum physics.

Albert Einstein was the central developer of both special and general relativity, and he also developed important aspects of quantum physics. *Time* magazine named Albert Einstein the man of the century.

Twentieth-century physics was important to the twentieth century.

Special Relativity

Lonnie is the stay-at-home type, but her twin sister Bonnie is an astronaut. At age 20, Lonnie starts to raise a family while Bonnie goes to visit the planets in a nearby star system.

It's a long trip for Bonnie and her crewmates, but this trip is taking place with advanced technology allowing travel at very high speeds. Great discoveries are made and, after the long voyage home, Bonnie returns just at her 30th birthday.

Lonnie is 70 years old and greets the returning astronauts at Cape Canaveral with her grandchildren.

Huh?

Lonnie is 70 and Bonnie is 30.

Moving clocks run slow.

This is a fact. A physical reality.

Whether Bonnie is 30 or 31 or 69 or 69.99999 years old when she returns—this is a detail that depends on Bonnie's exact speed on her voyage.

But Bonnie will be younger than Lonnie when she returns. All contemporary scientists agree with this reality. It's how the universe works.

Why we don't notice in our everyday lives that this is how the universe works is because we on earth travel very slowly.

Bonnie would have to travel at over 650 million miles per hour in order for the age difference to be 40 years. She would have to travel over 90 million miles per hour for there to be even a 1% difference in aging.

But just the concept—the idea that Bonnie's speed has anything to do with how fast time travels for her—sounds absurd.[4]

Perhaps more absurd-sounding is the idea that time is not universal. Time does not flow like a smooth river, second by second, the same everywhere. Time is a local phenomenon whose passage varies with how fast we travel.

There is nothing about this that matches anything from how we experience our everyday lives, but nevertheless it's true. It's completely imperceptible to us because we don't experience speeds of 650 million or even 90 million miles per hour, so we have no intuition for this. We may safely be ignorant of this fact, yet function perfectly normally for our entire lives.

Part of Einstein's brilliance is evidenced by his being able to draw these conclusions about special relativity simply by thinking about the implications of the speed of light being the same for all travelers. When I'm moving away from an object, the light waves emitted from this object will inevitably be more spread out for me than for someone not moving away from the object, due to my motion away from the object. But because the speed of light is invariable, it is time itself that must spread out.

There's more to special relativity, of course, and you don't get college credit for reading these few pages. For example, in addition to time passing more slowly for a fast-moving traveler, length contracts and mass increases. But for our brief overview, we'll just accept that, if our normal life took place at 650 million miles per hour, or if we routinely dealt with distances of intergalactic magnitude, we would never have created for ourselves the conceptualizations that we have of distance, time, space, and mass. Our everyday concepts don't work for scientists who work with the fast-moving particles of the subatomic world, and they don't work for scientists dealing with the vast distances of the universe. For these scientists, the adjustments developed through special relativity must be made, because special relativity is reality.

Concepts to Speculate On

One of the important aspects of special relativity that we'll be bringing forward to the rest of this book is how special relativity challenges our notions of the space and time dimensions.

We do not have a well-developed intuition for these dimensions.

Our size, the distances at which we operate, the timeframes within which we act, what we're built to sense and feel—these all conspire to create for us a naive, inadequate, incorrect view of the space and time dimensions in which we operate.

We're all in awe of the great magnitudes of space. But before special relativity, at least we felt that we could be masters of space—that the intuitions we are born with and have developed from our everyday lives regarding distances and speeds and time—that these intuitions are straightforwardly extendable to the far reaches of the universe. They're not.

The universe is so immense that it's understandable only in terms of light years, and only in conceptualizations that require us to think very large, very far, very fast, for very long periods of time. There's no other way to think about the universe. And when we think this way, our thinking must be adjusted to take into account special relativity. Otherwise everything we expect—all of our measurements of space and of time—will be wrong.

And these are our most familiar dimensions.

General Relativity

It gets weirder.

If you think of planets as giant masses revolving around the mass of the sun, held in place by gravitational force, you've got it wrong.

What's actually happening is that the sun reshapes space. Gravitational waves emanate from the sun and change the shape of space around it. For millions of miles around the sun, space is not shaped with a north-south dimension at ninety degrees to an east-west dimension at ninety degrees to a vertical dimension. Space is curved. And the planets float effortlessly, taking the path of least resistance through this curved space.

Now this may actually sound to you like a distinction without a difference. After all, what's the practical difference between a universe in which planets are held in place by gravitational force, and a universe in which the force of gravity is propelled at light speed via gravitational waves which reshape space?

Not much, it turns out. But there are differences, and they were first observed by precise measurements of perturbations in the orbit of the planet Mercury about the sun, and by similarly precise measurements involving phenomena that can be observed only during a solar eclipse, when measurements matched general relativity's predictions for the bending of light. Only after many decades of additional observation has near-unanimity been reached on the existence of gravitational waves.

Again, no college credit for this summary description. But the important point for this book's perspective is that—to understand the universe—we have more strange and counterintuitive concepts to absorb, for example a concept that the force of gravity is propelled in waves throughout the universe, reshaping space and time, and setting up the motions and interactions among the planets, the galaxies, all matter.

This understanding will not help you in your everyday life. In fact, your day will go just fine if you don't make any adjustments at all for either general relativity or special relativity. This is because we obtain only a very refined degree of additional accuracy by introducing relativity's corrections: Isaac Newton's seventeenth-century classical physics is good enough to point you in the right direction to get to work or to the grocery store, and you will not have to be concerned with the very small discrepancy that has been created between your wristwatch and those of the people you encounter. But relativity's adjustments do make our measurements of time and space just a bit more accurate, and perhaps more importantly they give us a truer understanding of how the universe works.

What's Going On With . . . Gravitational Waves

Before Einstein, gravity was assumed to propagate instantaneously. But general relativity changed this: gravity propagates in gravitational waves, which spread out at the speed of light—fast, but not instantaneous.

By the mid-1980s, it had become fairly certain that gravitational waves exist, although only by indirect observation of astronomical phenomena, such as examination of the orbital decay of a binary pulsar system. The effects of gravitational waves in space have also been observed as two neutron stars spiral towards each other, but there has never been a human-made instrument that has actually detected gravitational waves. So far, we've only inferred their existence from observation of celestial bodies.[5]

In order to have a chance at detectability by any of today's instruments, a gravitational wave would have to originate at some powerful, violent event, perhaps in the vicinity of a black hole or a neutron star. But, due to the weakness of the gravitational force, all detection devices available to date have trouble distinguishing a valid gravitational wave from the background effects that inevitably interfere. This is in spite of the elaborate arrangement of five parallel detectors set up worldwide, with each of these suspended detectors typically several tons in weight, supercooled to just above absolute zero, and equipped with superconducting quantum interference devices—SQUIDs—and with seismic damping devices.[6]

Enhanced equipment is under development and the work continues.

Unifying the Forces

What's key about this for this book is how much focus has been placed during recent decades on the force of gravity. Throughout the twentieth century, physicists have been obsessed with the creation of a unified force theory, and it is gravity that has proven the most troublesome force to unify with the other forces.

Four forces exist in the universe—the electromagnetic force, the weak and strong forces that operate within the structure of atoms, and the force of gravity.

Back in the nineteenth century, physicists had an even earlier success at this, unifying the force of electricity and the force of magnetism, by showing that these two forces are actually the "same" force, the electromagnetic force.

Now this is odd, because common sense—our common basis for understanding the universe—tells us that electricity and magnetism are not the same thing. They're different. One goes through wires, and the other involves magnets and iron filings.

But the point is that we're just not understanding the universe correctly if we continue to think that electricity and magnetism are different forces.

The physicist's perspective is that electricity and magnetism are best understood as dual aspects of a single phenomenon, modeled mathematically as interacting fields, moving in tandem, electricity producing magnetism, and magnetism producing electricity.

With electricity and magnetism unified, physicsts focused on unifying electromagnetism and gravity. But then two additional forces were identified within the physics of atomic particles—the strong force and the weak force—increasing from two to four the number of basic interactions of the universe that physicists are now challenged to unify.

So why the obsession? What's so important about unifying the forces, creating a physical framework in which there is only one force?

The answer is the big bang.

If the universe began as a miniscule point that has been expanding for fourteen billion years—well, what was in that miniscule point? There couldn't have been room in there for four forces (so the thinking goes, loosely expressed). There must have been just one force, whose manifestations varied (especially as experienced through our simple senses) as the universe expanded (and cooled). So let's figure out how it can be that these four forces can all be understood as one. Let's move backward in time to the big bang.

Now you may not be convinced that the only conclusion we can draw from the fact of the big bang is that there exists only one, unified force. You may not even be convinced of the fact of the big bang. But the fact is that, as the twentieth century ended, physicists had come very close to achieving a consensus view—within the standard model of particle physics—on how it is that all the forces except gravity are unified. This consensus view is called the Grand Unified Theory, and we'll be discussing in more depth the mainstream view of the extent of Grand Unification. It's certainly been a great achievement of physics and mathematics to have gotten so far during the past century.

You may be thinking: aren't physicists getting a bit carried away here, in calling Grand Unification something that seems more properly called "Three-Quarters Unification"? After all, "Grand Unification" is unifying only three of the four known forces.

Maybe you're right if this is what you're thinking. But if so, physicists have paid a price for this overstatement: what will they call it when gravity too is brought into the unification? Their answer to that is the "Theory of Everything." That's what physicists are striving for as they work to unify all of the forces, including gravity.

We'll have a lot more to say about Everything. But first we have the third and strangest strand of twentieth-century physics.

Quantum Physics

For much of the twentieth century, physics has focused on quantum physics, the physics of the very small, the physics operating inside the atom.

The thrust of quantum physics is that physical phenomena are not continuous phenomena, but instead take place in very small but discrete increments—that is, quanta. These quanta cannot be isolated with certainty or specificity, but instead we can say only where they're likely to be. In fact, quantum physics—through the Heisenberg uncertainty principle—tells us that there are limits on how much we can know: the more accurately we can locate a quantum particle's position, the less we are able to know about another core characteristic of the particle, its momentum.

Quantum physics is incredibly accurate, and not just about the limitations of our knowledge. The equations that probabalistically point to the physical world's quanta have given us another great leap in our accuracy of measurement of space, time, matter, and other physical phenomena. Because of quantum physics' accuracy, we have made great advances in chemistry, biology, electronics, and other applied sciences: lasers, bar code readers, compact disc players, global positioning devices, DNA, genetic engineering, microsurgery, semiconductors, nuclear energy, superconductors, spectral lines, the stability of the atom—all have applications that are due to the great accuracy of the equations of quantum physics.

Quantum physics is also about subatomic particles—identifying and ultimately observing the particles that make up matter. And also the particles—gravitons, photons, gluons, W and Z bosons—that transmit the four forces. (Most physicists would probably prefer here the terminology: gluons are the exchange particles of the strong interaction, W and Z bosons are the exchange particles of the weak interaction, and so on. But in this book we'll tend to use slightly less technical language—transmit the forces—where we can.)

There is another strand of quantum physics, with numerous interpretations, some philosophical, perhaps even mystical. Quantum physics stretches our understanding of reality. It is, in some interpretations, filled with dualities and contradictions. Matter emerges from then disappears into a great quantum vacuum. Particles can't both be and be known to be. Matter shifts from existing to only having the potential to exist. The act of measurement distorts what's being measured. Human consciousness seeps into the discussion of quantum physics. Our human acts affect what is true at the quantum level.

With coming up to a century of effort, much has been done to demystify quantum physics. It's not all dualities and contradictions, and in fact these won't be useful concepts for us to proceed with. For this book, we'll need to know that quantum physics has important deterministic elements and important nondeterministic elements.

In the mathematics of quantum physics, the state function—the formula for how the state of a quantum particle evolves over time—

changes in two different ways. Deterministicaly, by continuous causal evolution, one step triggers a next step triggers a next step. This is in contrast to the second way in which the state function can change, the nondeterministic element of quantum physics—collapse at measurement.

In quantum physics, multiple possibilities exist—are superposed—and it is not until observation (or, synonymously for this purpose, measurement) that the state function collapses to just one of these actualities. Without observation or measurement, the superposed (multiple) possibilities proceed in parallel, but deterministically.

Parallel realities are not reduced to a single path until observation triggers a quantum jump. This is without question bizarre. It seems far distant from a scientific view of the world and leads to basic questions, which we'll touch on in Chapter 5, about the nature of reality.

Quantum physics' nondeterministic elements are probabalistic (Einstein referred to them, dismissively, as God throwing dice): when quantum physics' probabalistic phenomena are observed, we don't know in advance what we're going to see; we know only what we might see and how likely each possibility is.

Quantum physics correctly predicts that, when the position of a particle is repeatedly measured or observed, we will not come up with the same answer every time, but instead will come up with a range of answers whose likelihood is predicted by the Schrödinger equation for the quantum wave function, a central equation of quantum physics. This amazing equation incorporates both the parallel deterministic time evolution of the superposed quantum possibilities, and also each quantum possibility's probability that it is how the quantum system will collapse upon observation.

There are a number of interpretations of quantum physics—interpretations of reality—which we will be discussing later on. Schrödinger himself did not accept that physical phenomena are not real until they are observed, and he developed the somewhat gory thought experiment—now referred to as Schrödinger's cat—to demonstrate the absurdity of Bohr's Copenhagen interpretation of quantum physics. Schrödinger's cat is unobserved in an enclosure in which the cat may or may not have died from poison gas, and it was Schrödinger's intention to show the absurdity of the cat's living or dying being indeterminate until the lid is opened and the cat observed. Schrödinger, with Einstein, rejected the Copenhagen interpretation's observer-dependent reality, and preferred a realist interpretation, in which the cat's fate is real even before it is observed.

When Is Classical Physics Good Enough?

For many purposes, classical physics is good enough. That is, our physics is accurate without introducing quantum corrections. After all, those high school physics tests had very precise answers to questions about blocks sliding down inclined planes and planets in motion.

There is no quantum indeterminancy in physics' macroscopic events and objects. Macroscopic events can be correctly described without worrying about the submicroscopic uncertainties of the quantum world. At the macroscopic level, unlikely quantum possibilities cancel each other out and events transpire according to classical Newtonian physics, with relativistic corrections, if needed, but without quantum corrections.

The dividing line between classical physics and quantum physics is traditionally drawn somewhere between macromolecules (complex molecules) on the classical side, and atoms, electrons, and photons on the quantum side.[7] So proteins and cells can be treated classically. And the brain, even at the level of the single neuron, is subject only to classical physics, not to quantum physics, in the view of most neurobiologists and other scientists. There are minority views on this subject among physicists, however, and we'll be discussing this whole topic in depth later in this book.

The central quantum phenomenon—the collapse of the wave function, or the quantum jump—occurs when we observe or measure a quantum process, resulting in the forcing of the quantum hand. A quantum system, when observed or measured, turns over its cards; it can no longer reside in a world in which it can be many things (each with its own probabiltity). When we observe or measure, the quantum card becomes specific.

This central quantum phenomenon has many names, many ways of labeling it and thinking about it. It is called the reduction or the collapse of the wave function. It is also called the quantum jump or the quantum leap or the quantum transition. It is quantum decoherence: the cohered probabilities, layered one on top of each other in quantum superposition, decohere to create classical properties. It is the migration from the sum over histories, in which all quantum possibilities accumulate, to a single classical event or particle. It is "popping the qwiff."[8]

What's Going On With . . . The Quantum Jump

The 1997 Quantum Future symposium had numerous discussions of the quantum jump. Many of these discussions reflect physicists' reductionist urges. Can we better understand the quantum jump by examining it more and more closely, even reducing the quantum jump into component parts? Can we identify and map out quantum microprocesses, the quantum jump's *chain of reduction*?[9]

It's only nonscientific superstition to believe that a watched classically scaled pot won't boil. But this is exactly what happens at quantum scales: repeated measurements prevent the natural evolution of a quantum system.[10]

Physicists propose an experimental set-up to visualize the quantum jump by continuous fuzzy measurements—measurements not obtrusive enough to immediately trigger a collapse of the wave function, but strong enough to meaningfully measure characteristics of the quantum transition in progress.[11]

Physicists try to relate the core phenomenon of quantum physics, the quantum jump, to other phenomena of physics, even classical physics. Some look at the physics of classically scaled irreversible processes and conclude that the quantum jump can be considered just an ordinary irreversible process.[12]

It bothers some physicists that the concept of the quantum jump has been "replaced by something even less appealing—a discontinuous element in the evolution over time which is no longer restricted to the scale of the single quantum event."[13] Unappealing or not, this very concept—that the quantum jump is a phenomenon at all scales—becomes core to the punchline of this book, many chapters in the future.

Physics' Contradicting Strands

At the deepest and smallest scales, quantum physics contradicts general relativity. General relativity's reshaping of spacetime through the force of gravity is a smooth reshaping. But the submicroscopic world of quantum physics is a cauldron of possibility: the quantum forcing of hand is discrete and discontinuous at the quanta of space and time.

Unlike general relativity's picture of a smooth and continuous reshaping of spacetime at all levels, there is nothing smooth about the quantum view of spacetime. Gravity again rears its head as the mysterious force: general relativity has permitted us to understand gravity's influence on the great masses and distances of space, but general relativity seems to contradict quantum physics when gravity is brought down to the scale of the smallest worlds of the elementary particles.

***

The next chapter is a review of the state of physicists' degree of acceptance of the current explanations for the phenomena of physics—a report card on twentieth-century physics. After that, we can explore further some of the more tentative theories and findings to see where they might lead.

2.
Report Card on Twentieth-Century Physics

Roger Penrose is a prominent and respected mathematician and physicist, a professor and researcher at Oxford University. Perhaps his prominence is part of the explanation for the sharp opposition he faced from many scientists to his 1989 book, *The Emperor's New Mind: Concerning Computers, Minds, and the Laws of Physics*. We'll be discussing later the conclusions Penrose presents with respect to how quantum physics relates to consciousness and the mind—it is these conclusions that have been sharply opposed by many scientists.

But this chapter is not about these conclusions of Penrose's book; it's about some stage-setting of Penrose, in which he provides an evaluation of physics' theories from throughout the ages. We'll use Penrose's analysis as our jumping-off point for establishing the end-of-twentieth-century consensus report card for modern physics.

This report card is largely a progress report on the unification of quantum physics with the rest of physics, and the unification of physics' forces with each other and with matter. This reflects physics' mainstream lines of twentieth-century inquiry.

***

Roger Penrose has gone out on a limb to characterize as Superb, Useful, or Tentative the theories of physics that have been developed over the centuries.[14]

Penrose is not one those easy graders, handing out A's to just about his whole class. In fact, for fields other than physics, Penrose mentions that he would be hard-pressed to find any Superb theories at all.

Evolution comes to Penrose's mind as a rare example of a non-physics Superb theory of science.

So physics is blessed in having six Superb theories, theories whose accuracy and breadth have provided great leaps forward in human understanding of the universe: Euclidean geometry, Newtonian mechanics, Maxwell's theories of electromagnetism, special and general relativity, quantum physics, and quantum electrodynamics.

Few mainstream physicists would dismiss as less than Superb any of these six theories, so let's make sure we understand how far they take us. Let's also take a step back from this selection of Superb theories in order to deduce the underlying assumptions about twentieth-century physics' central questions and directions.

Maxwell's theories—developed standing on the shoulders of the giants Newton and Euclid before him—are the nineteenth-century accomplishments that we've already mentioned unify our understanding of electricity and magnetism. Relativity and quantum theory have greatly improved the accuracy of our understanding of the world. They have also whetted our appetite for unifying the four forces, but at the same time they've upped the ante for how accurate our understanding needs to be.

The quantum electrodynamics (QED) theories have greatly improved our understanding of one of these forces—electromagnetism. QED incorporates special relativity and quantum physics into Maxwell's equations in order to allow a virtually complete understanding of the electron and the electromagnetic force.

Penrose notes, as do other physicists, that even QED is not quite completely understood. But Penrose nevertheless is willing to concede QED as Superb. His mild reluctance in this regard probably places Penrose at the somewhat conservative end of the spectrum of today's physicists, as does Penrose's relegation to merely Useful three theories that other physicists may be more fully comfortable assessing as Superb: the big bang, the electroweak reconciliation, and quantum chromodynamics.

The big bang is at its heart a theory of cosmology, of how the universe of today has emerged over the past fourteen billion years. And we've already mentioned that it provides impetus for a driving goal of today's physicists, the goal of unifying all of the forces.

Particle physicists—the physicists of the small—also look to the big bang, as a framework for investigating the elementary particles of matter. This is because moving backward in time toward the point of the big bang, we're moving not only smaller, but also toward enormously higher temperatures and energies. The high-energy physics of particle physicists' accelerators allows experimental physicists to mimic the environment of the smallest fractions of a second after the big bang, and as a result to see what it is that we're really made of. In fact, the Theory of Everything that physicists are aiming to complete is a theory that unifies both forces and matter—it is a theory, once created, that puts forth a single understanding of every force and every elementary particle of nature. The big bang is certainly a Useful framework for physicists.

The electroweak reconciliation generalizes Maxwell's equations of electromagnetism to also incorporate a second force, the weak force, which is responsible for radioactive decay. And quantum chromodynamics (QCD) parallels QED to create an understandng of the strong force that operates within the atomic nucleus, and an understanding of the quarks that the particles within the atom's nucleus are made of.

The selection, development, and understanding of these Superb and Useful theories reflect twentieth-century physics' focuses on unifying all of the forces and matter, and on bringing quantum physics into this unification. But these Superb and Useful theories together do not yet unify even the three nongravitational forces: QCD explains the strong force, but does not "unify" it with the electroweak. Penrose places this Grand Unification Theory—unifying the electroweak and strong forces—as one of his four Tentative theories.

At this point, we can stop for a moment to assess again whether Penrose's grades are too harsh—whether today's Grand Unification Theory is not established sufficiently to be accepted as the consensus unification of the three nongravitational forces. There are certainly other physicists who are comfortable with accepting that there's just one more force to unify, and who have moved on to incorporate gravity into a Theory of Everything. But there are also physicists looking to dot all i's and cross all t's before saying that we know everything about the electromagnetic, weak, and strong forces and how they are unified into one.

And there are also physicists that are going for the brass ring—the Theory of Everything. Penrose's remaining three Tentative theories are three components of a proposed Theory of Everything: string theories; supersymmetry (nicknamed "SUSY"), which goes beyond the unification of the forces themselves to also unify the force-propelling particles with the particles of matter; and theories of extra dimensions (called the Kaluza-Klein theories), which create a topology that unifies the forces in dimensions beyond our familiar three spatial dimensions plus time.

These Tentative theories have many adherents, and are subjects of great current focus among many of today's physicists. A theory called M

Theory combines the various string theories with these other Tentative theories and will be discussed in the next chapter.

But these theories also have their detractors, Penrose among them. Penrose mentions that he resisted the temptation to split his Tentative category into a fourth category—Misguided.

Nevertheless, a discussion of twentieth- and early-twenty-first-century physics must include these possibilities for achieving a Theory of Everything, and must note twentieth-century physics' march toward the unification of the forces and matter, and the reconciliation of relativity with quantum physics.

Nobel Prizes for Physics

Awards of the Nobel Prize for physics track fairly well with Penrose's modern Superb theories plus his Useful theories. These are key Nobel Prize awards that recognized advancement in the understanding of quantum physics, relativity, and physics' fundamental forces:

Max Planck (1918) for his discovery of energy quanta

Albert Einstein (1921) for his "services to theoretical physics," recognizing in particular his work on the photoelectric effect, which is a key quantum phenomenon

Niels Bohr (1922) for his investigation of the structure of atoms

Louis-Victor de Broglie (1929) for his discovery of the wave nature of electrons

Werner Heisenberg (1932) for the creation of quantum mechanics

Erwin Schrödinger and Paul Dirac (1933) for new discoveries in atomic theory

Wolfgang Pauli (1945) for the discovery of quantum physics' exclusion principle

Max Born (1954) for the statistical interpretation of the quantum wave function

Sin-Itiro Tomonaga, Julian Schwinger, and Richard Feynman (1965) for their work on quantum electrodynamics

Sheldon Glashow, Abdus Salam, and Steven Weinberg (1979) for the electroweak unification

Carlo Rubbia and Simon van der Meer (1984) for work on the particles transmitting physics' weak force

Jerome Friedman, Henry Kendall, and Richard Taylor (1990) for work leading to the development of the quark model of particle physics

Gerardus 't Hooft and Martinus Veltman (1999) for the quantum physics of the unified electroweak force

This selection of Nobel Prize winners is intended to show the march of twentieth-century theoretical physics toward a unified, quantum understanding of physics' forces. This march also underlies Penrose's selection of Superb and Useful theories.

Other Nobel Prizes for physics have been awarded over the years for developments in elementary and atomic particles and their interactions, in our understanding of the cosmos, and in the tools of experimental physics. And we'll also, in the course of this book, be meeting additional winners of the Nobel Prize, for their work in some of the phenomena of new physics.

***

The list of Nobel Prize winners, like Penrose's nominations for Superb and other theories, is in large part a report card on the progress toward achieving a Theory of Everything, which is defined as a theory that reconciles relativity and quantum physics, and that creates a single understanding of all of physics' matter and forces. The Nobel list stops before even three of physics' four forces are unified, at Grand Unification, which is one of Penrose's Tentative theories. And Nobel awards fall far short of a Theory of Everything, which unifies all of physics' forces and for which Penrose selects some elements as Tentative.

Although string theory has not yet earned any physicist a Nobel Prize, it is today's most widely advocated proposal for a Theory of Everything.

3.
Entering the Twenty-first Century:
String Theories

Many physicists accept string theory as a framework for solving questions that remained open as the twentieth century closed— unifying gravity with the other forces, reconciling general relativity with quantum physics, explaining the elementary particles. String theories have now been combined with the concept of "supersymmetry" to create a framework that moves us toward a Theory of Everything.

String theory is a vast subject for which only the briefest summary is offered here. It is so vast a subject because it is today's most widely explored theory of modern theoretical physics. Whole conferences are held to discuss aspects of string theory, advanced applications, and new developments. It is undoubtedly a powerful analytic tool, under continual refinement by physicists worldwide who strive to perfect string theory in order to explain all of the physical world.

***

The "strings" referred to within string theory are theoretical vibrating entities that are the smallest building blocks of subatomic matter. They present a framework to explain what otherwise seem to be the completely arbitrary relative magnitudes of the four forces, the particles that transmit these forces, and the building blocks of matter at the subatomic level. The natural vibrations of these strings explain how all this comes together: as these strings resonate within the topology of extradimensional space, their harmonic patterns create all of physics' elementary mass and force particles.

At the heart of string theory is the concept that space and time are granular, that there are a smallest distance and a smallest time—

the Planck length and the Planck time. The granules are very small (distance granules of 10^{-33} centimeters, and time granules of 10^{-41} seconds, if you're counting).

Because "spacetime" is granular, there is a limit to the extent that gravity reshapes space. This permits a reconciliation of general relativity with quantum physics, avoiding inconsistencies that would otherwise exist if spacetime shrank to a dimensionless point. And this permits strings, whose size is of the magnitude of spacetime's granules, to be the unifying entity creating all of physics' forces and matter.

This powerful framework requires the existence of ten (or so) dimensions. It's only within these extra dimensions that the mathematics and physics work out, that we can construct a set of string vibrations that can produce the patterns of mass and size and other attributes observed for all of the elementary matter and force particles.

Strings are proposed as physics' unifying concept. Their vibrations— in extradimensional space—unify all of physics' elementary particles and forces.

It's hard to explain extra dimensions, by drawing on paper or even by modeling in three dimensions, in a way that makes them accessibly understood. Mathematicians and physicists tend to prefer to model extra dimensions within mathematical frameworks that offer powerful explanatory capabilities. The next chapter will present another way to model and understand extra dimensions.

The original development of string theory was in the context of twenty-six dimensions. Later applications built a consensus around ten dimensions (our familiar three spatial dimensions plus time plus six more spatial dimensions). There was also debate as to whether the ten-dimensional framework is really nine. And today's latest thinking is that there may be eleven dimensions.

At this point in theoretical physics, there is not consensus that string theory and extradimensionality create a valid theory that works to explain the problems at hand. Nor is there consensus as to how many dimensions it implies. Nor is there consensus that this theory—even if true—is a basic theory, rather than one that just leads us to more questions and ultimately an even deeper and more penetrating theory. Even Michael Green, one of string theory's prominent theorists, emphasizes that these extra dimensions are only a possible prediction of string theory.[15]

The extraordinarily complex mathematics has not yet been completed to allow theoretical physicists to connect string theories with the realities that experimental physicists are able to observe as the specific nature and size of the matter and force particles. Some theorists explain this incompleteness this way: "The problem is that while twenty-first-century physics fell accidentally into the twentieth century, twenty-first-century mathematics hasn't been invented yet."[16]

Experimental physicists, among others, can be rubbed the wrong way by an explanation such as this, an explanation that relies on a claim to century-leaping advancement in theoretical physics, unmatched by other branches of science. There's a bit of a rivalry—often but not always good-natured—between experimental and theoretical physicists that in its form is pretty predictable: theoreticians are glad to see the big picture rather than drown in the mundane; experimental physicists can feel that they're doing the real work of physics, building and adapting particle accelerators and other tools of advanced physics to obtain concrete evidence, not theoretical constructs. Nobel-Prize winner and Harvard physicist Sheldon Glashow, for example, expresses the experimentalist's lament: "The superstring vision . . . requires the highest inaccessible dream-like energies to build a theory that deals with the down-to-earth world under our feet."[17]

But generally physicists are at least intrigued—many are much more enthusiatic—by how variations on string theory have been combined with the concept of supersymmetry to create the potential for an internally consistent framework that is a Theory of Everything.

Supersymmetry is the concept that, at high enough energies, there is a partnering between every matter particle and a sibling force particle, and vice versa. The supersymmetric siblings have never been found experimentally, but what's exciting about this concept is that—if found—they would take us a long way toward unifying the elementary particles of matter and the elementary particles of force, which is a task that a Theory of Everything must achieve in addition to unifying the four forces into one. For example, supersymmetry posits a supergravity matter sibling, the gravitino, to the gravity-transmitting graviton.

As physicists thought more about string theory and supersymmetry, an embarrassment of riches developed—five string or "superstring" theories. More work—plus an eleventh dimension—convinced physicists that these theories are all consistent with an umbrella theory now called M Theory. There's some uncertainty about what the M stands for—it could be Mystery, or Magic, or Membrane (that is, a multidimensional string), or Matrix, or maybe Mother of all theories. Today the M Theory generalization of string theory is the most highly accepted framework for academic physics, even if its detractors—such as those suggesting that the M stands for Murky[18]—haven't yet given up.

A number of physicists are exploring extradimensional space as a key concept, even beyond its use within M theory. This is somewhat ironic, in that string theorists tend to view the extra dimensions as a by-product of the more core concept of strings.

It strikes some physicists that the M Theory of Everything is digging itself deeper and deeper into mathematical modeling that can't yet be accomplished, and into hypothetical structures and particles that are so minute and ethereal as to offer no chance at reasonable near-term testability.

Strings undoubtedly present an accurate view and a unifying theme. But a question to be asked is: View of what? Is enough being unified, if string theory or M theory doesn't extend to the full scope of quantum phenomena, or extend to some of the important concepts of new physics discussed in Part Three, or extend to consciousness and the mind?

It is the premise of this book that string theory suffers from two false steps. It is overly reductionistic, looking for The Answer by dissecting more and more finely, rather than seeking a holistic view. And it has extrapolated from the past overly linearly, seeking The Answer in the resolution of historical conflicts, in the unification of general relativity and quantum physics, the unification of the forces, the unification of matter and force.

It is the premise of *New Physics and the Mind*, along with some of today's radical theoretical physicists, that an alternate route to a true Theory of Everything may be available through a holistic view of physics, which brings in consciousness and the mind, and which incorporates some surprising developments of new physics.

These alternate routes will be discussed in upcoming chapters, along with a very basic challenge to a whole premise of string theory's unification of gravity with the other forces. String theory's focus on unifying gravity with the other forces does not take into account gravitational exceptionalism: only gravity reshapes space. The other forces operate against the background of spacetime, but in a very real sense general relativity tells us that gravity is the background of spacetime.

We'll revisit this whole question in Chapter 12 (Quantum Gravity). Many physicists view gravity's unification with the other forces as the critical extension of the twentieth century's quest for force and matter unification as a theory of everything. But other theorists are skeptical that a theory can be the basic theory of everything if it attempts to unify the un-unifiable by using the same concept to explain both background-independent gravity and the other forces, which are background-dependent.

Compressed Extra Dimensions?

After the big bang, so the theory goes, forces and matter expanded to create the universe that we live in, a universe that is continuing to expand. The string theory hypothesis with respect to the extra dimensions is that these extra dimensions (that is, dimensions #5 and higher) remained extremely compressed.

Don't forget: long ago, physicists began viewing time as our fourth dimension. We'll come back to this later in this book—in fact, we'll need to discuss the arrow of time and imaginary time, and we'll also be asking if time really exists at all. In the meantime, it will be convenient to accept the established conventions of dimension numbering: #1 through #3 are the familiar spatial dimensions of our 3-D world, and #4 is time. So extra spatial dimensions begin at #5.

The compression of these extra dimensions is at the heart of much of string theory. And it is only because these extra dimensions are much smaller than our smallest observable lengths that they work as tools of explanatory power: they're too small for us to have noticed them, even with the latest sophisticated tools of measurement and observation.

Within the framework of the big bang model, a small extradimensional universe with no specially compressed dimensions was the earliest universe. For the briefest moment after the big bang, all of the forces were unified, and in addition force and matter were the same thing. The search for a Theory of Everything is a search for a theory that reconstructs this unified moment. This was the briefest and most fleeting moment, an extraordinarily tiny world of extraordinarily high temperature and energy. According to the string extradimensional framework, it was also a world of ten (or so) dimensions of space.

Using the terminology of physics, as the universe expanded, the first symmetries of this world that broke were the symmetries unifying matter and force, as well as the symmetries unifying gravity with the other forces. As these symmetries broke, the extra dimensions remained compact while our three observable spatial dimensions plus time continued to expand.

Generally, extradimensional space has been viewed as a mathematical tool to make sense of constructs such as the unification of the forces, or strings, or supersymmetry. Some physicists have wondered if the fundamental universe might have just two dimensions,[19] or if the extra dimensions are in fact not dimensions of space.

There is now a significant effort among physicists to take seriously these extra dimensions as concrete and real dimensions of space. Experimental physicists are working to find physical evidence of these extra dimensions—to map out these extra dimensions and the shadow matter that they may contain. Some physicists propose that the nongravitational forces operate only in our familiar three spatial dimensions plus time, but that gravity—the weakest force in our dimensions—is more fully expressed in dimensions #5 and higher.

***

We'll be going into depth on the latest research and theories of extradimensional space. But first we need to spend some time bringing along the physics of consciousness. To move in this direction, we'll start with a discussion of the competing strands of quantum physics' philosophical framework.

That sounds like such a heavy topic. Let's have a brief, lighter digression into how extra dimensions might be experienced by us, the human being.

4.
Ghosts

Some of Part One, Entering the Twenty-first Century, has certainly been consistent with the concept of truth being stranger than fiction. So here's an admittedly fictional interlude, a bit of arts and entertainment. It's not meant to (necessarily) be taken literally. But hopefully it will be fun and will give us a bit of practice trying to visualize higher-dimensional space.

***

Anyone who has read the wonderful nineteenth-century book *Flatland*[20] should have an easy time with this chapter.

Flatland envisions a world with only two dimensions—a flat land. Two-dimensional beings, with shapes such as circles and triangles, float across this flat land, entering and exiting their pentagon houses through breaks in the pentagons' sides. It's a happy world of north/south and east/west. Until one day a sphere (three-dimensional!) crosses Flatland's plane.

What's clear to most Flatlanders is that ghosts have arrived. For what they see, whenever the sphere passes through the plane of Flatland, is a point that expands smoothly to a larger and larger circle, then starts shrinking to a smaller and smaller circle, then is reduced to a point, then disappears.

Clearly a ghost within Flatland.

The three-dimensional sphere, passing through two-dimensional Flatland, is experienced as an eerie spectre that appears from nowhere, fades in, fades out, then disappears.

An extradimensional object, traversing our three-dimensional world, could be experienced as this ghost-like shadow, entering, exiting our world.

An altogether plausible understanding of how ten dimensions have been observed?

5.
Entering the Twenty-first Century: What Is Reality?

Quantum physics, from its earliest constructions, requires all scientists to formulate a stance about the nature of reality. This is because of the quantum jump, the jump from unknowable multiple paths to a known single path. Scientists' understandings of quantum reality range from excluding from science any role for the mind to embracing the quantum jump as science's moment of consciousness.

***

Niels Bohr and Albert Einstein are early proponents of two distant ends of the spectrum of theories of quantum reality.

Bohr and the associated Copenhagen interpretations of reality emphasize reality's subjective nature. There are shades of Copenhagen interpretations, in which deep reality is created by observation, or deep reality does not even exist. Physicist Fred Alan Wolf, whose work is discussed in Chapter 11, is in the school in which reality is created by observation.[21]

In other shades of Copenhagen, it is consciousness, rather than observation, that creates reality. Proponents of this view include, John von Neumann, mathematician and father of modern computer science, as well as Henry Stapp and Eugene Wigner, whose work we'll be discussing later.

Beyond the Copenhagen interpretations, we find proposed understandings that link quantum physics with ancient mystic thought and understandings of existence. Capra's *The Tao of Physics* emphasizes the yin and yang, the undivided wholeness, of quantum physics.

Einstein and the realists, by contrast, emphasize the ordinariness of the processes of physics. Realists insist that nothing about quantum theory requires that we introduce into physics anything but concrete physical objects and interactions. Anything else is mystification, and there is no reason to bring the mind or consciousness into even the deepest understanding of physics.

And we find understandings that seem to lie between the poles of Bohr and Einstein. The many-worlds interpretation emphasizes parallel universes that develop as quantum processes, proceeding in parallel, unnoticed until observation or measurement creates branches and sub-branches of parallel quantum realities.

Perhaps most striking about these various world views is that they're views of physicists and mathematicians, not purely the province of philosophers.

Nick Herbert, after his analysis of the full range of understandings of quantum reality that have been put forth, can't avoid asking: "Is consciousness a type of quantum knowledge?" He comments: "Science's biggest mystery is the nature of consciousness." And he asks: "Is it possible that consciousness is some sort of quantum effect? Is human awareness a privileged access to the 'inside' of the quantum world, an open door to some brain quon's [elementary particle's] realm of possibility?"[22]

***

Previewing what's ahead in this book, it's hard to reject any of physics' great thinkers' conclusions about reality. The debate goes back to the beginnings of quantum physics, and if anything has gotten more complex as surprising discoveries of new physics unfold.

Is synthesis possible? Can a unifying vision be created from these apparently contradictory views?

I would say yes. And I would say this has been accomplished in our best theories of new physics. Being able to accomplish this synthesis will be, I'd say, the mark of the great twenty-first-century theory.

To accomplish this, the task will be to create a universe of reality from observation and from consciousness. And to permit many universes. And to do this through an undivided wholeness. And to understand consciousness as a phenomenon of physics. And, at the same time, to keep the world as one made of ordinary objects.

Quite a challenge, as these views seem largely mutually exclusive.

And theories of the nature of reality are not the only sphere in which we'll be challenged to incorporate different but each highly accurate visions of how the physical world works. Even different highly accurate visions each proposing powerfully unifying concepts.

String theories, for example, have created an exceptionally accurate physical model with a bare minimum of basic concepts. And a number of other theories we'll be discussing later also create accurate and streamlined physical models.

What do we do with all of these riches?

How many unifying concepts will we allow? And are all unifying concepts created equal?

No. Some are more equal than others. And the best will be those that devour new observations, not those that new observations devour.

I'm proposing as the test for the best theories of new physics those that embrace a wider and wider scope of phenomena and gain strength through this embrace. Not theories which incorporate a wider scope of phenomena only by growing additional limbs, additional crutches. Or theories that are a crutch. Or theories that substitute one crutch for another.

Admittedly, there's a bit of Occam's razor here. William of Occam was a fourteenth-century philosopher who is still invoked today to help with decision-making. The Occam's razor principle is that the simplest explanation is the most likely to be correct. Some of our theories of new physics will sound quite the opposite of "simple," but remember: we've got some very large, unconnected puzzles lying around: What is quantum reality? How do we reconcile quantum physics with general relativity? How do we unify matter, gravity, and the other forces? How do we explain the observations of new physics? What is consciousness?

As we discuss consciousness and the mind, and as we discuss scientific puzzles of late-twentieth-century physics, perhaps you too will be wondering: is this all of one piece? And perhaps you too will hear Occam whispering: Use the razor . . .

We'll keep this in mind when we look at speculative theories of new physics in Part Four. We'll earn the right to speculate after more science in Parts Two and Three.

But first, some final comments on an underlying assumption of twentieth-century science's phiosophical foundation.

6.
Reductionism, and Its Critics

String theories intersect with questions of reality at reductionism.

String theories look to answer the most basic questions of physics by breaking them down to a smallest basic part.

This is similar to much of twentieth-century physics' quest for the truly elementary particle and for the elemental intersection—at quantum gravity—of quantum physics and general relativity.

The success has been inarguably rapid, but there is an open question of whether mid-course correction is now needed toward holistic perspectives.

***

Reductionism is the principle that the whole is best understood by analyzing its parts and how these parts interact.

Ultimately, reductionism gives us a complete understanding of the universe at the theory of everything, when we understand that everything obeys the same fundamental laws: all of matter is unified, all of the forces (interactions) are unified, and this unification extends even to a single understanding that incorporates both matter and force.

Reductionism developed in part as an alternative to Descartes' dualist view of the world, in which the nonmaterial world of the mind is a separate (dual) world, distinct from the physical, material world of matter, space, and time.

In contrast to dualism, reductionism in its most complete form looks to explain consciousness, to explain life, as purely physical phenomena, without an appeal to phenomena residing in a second, nonphysical, nonmaterial world. Reductionism reduces the mind to the brain and its chemistry, and reduces life to the chemistry of DNA and cell biology. Reductionism eliminates the need for Descartes' nonmaterialist world: this nonmaterialist world can be reduced to material parts and their interactions.

The modern critics of reductionism do not return to dualism. Instead they look at consciousness, and at life, as phenomena that emerge from their component parts with their own levels of natural understandings, not understandings that derive from the operations of their component parts. Complex phenomena are different qualitatively, not just quantitatively or incrementally, from their less complex components. A relational holism overlays all of the parts, at all of their interrelating levels. The whole is not just the sum of its parts, and these holists de-emphasize elementary particles and elementary forces as the explainers of everything.[23]

Perhaps ironically, some now worry that holism has been so successfully argued in some circles that it has become a new orthodoxy, a victim of its own success. Some worry about the "advent of a new postmodernist fashion," a fashion that draws from principles of quantum physics—the role of the observer, wave/particle dualism, uncertaintainties of knowing—and extends these principles to academic analyses in fields of history, sociology, politics, and philosophy.[24] So the critics of reductionism are not without their own critics: "There is something awry about a theory that has exerted such widespread influence while effectively raising incomprehension to a high point of orthodox principle."[25]

We will find many philosophical routes on our road to new physics and the mind.

***

"Goodbye to Reductionism" was one of the presentations at a 1998 conference "Toward a Science of Consciousness."[26] And it is within the quest for a science of consciousness that materialist reductionism may be particularly inadequate.

But physicists did not immediately jump into the exploration of consciousness—certainly a radical field of scientific exploration—with the second radicalism of holism. The theories of Part Two's Physics and the Mind largely reflect a philosophy of reductionism. It was bold enough to suggest that physics has any role to play in a discussion of the mind.

Part Two
Physics and the Mind

What is the mind? What is consciousness?

Are these subjects properly part of science?

Philosophers view the mind as a function of the brain, but physicists have not uniformly embraced consciousness as a part of science.

Some do, however, and we'll discuss physics and the mind as Part Two of *New Physics and the Mind*'s four parts.

Part Two's arts & entertainment section: Doris Lessing.

***

Part Two Summary

Part Two is a review of the work of a small number of twentieth-century physicists who have proposed links between modern physics and the mind.

Chapter 7. Consciousness. The modern view of consciousness as understood by philosophers, psychologists, and other scientists.

Chapter 8. Early Extensions of Quantum Physics into the Social Sphere. Quantum physics, as well as relativity, have offered politicians and others a tool of metaphor, through which they have promoted various positions over the years.

Chapter 9. Doris Lessing. A brief digression to the career of the great twentieth-century writer, who partway through her career migrated from a fiercely realistic style to frankly speculative excursions into the mind and elsewhere. Reactions by Lessing's fans parallel the anger, disappointment, and befuddlement that have met physicists who propose scientific theories of consciousness.

Chapter 10. The Emperor's New Mind. The 1989 publication of *The Emperor's New Mind: Concerning Computers, Minds, and the Laws of Physics,* by respected mathematician and physicist Roger Penrose, is a seminal event for scientific inquiry into the mind. Penrose's groundbreaking proposals include a link between the mind and a central question of modern physics, which is how to reconcile twentieth-century physics' two major strands, relativity and quantum physics. Penrose also proposes biological mechanisms that take advantage of this link. First reactions to Penrose were largely unsupportive.

Chapter 11. More Physicists' Thoughts about the Mind. Both before and after Penrose, other physicists have proposed how modern physics might help explain consciousness and the mind. Some of the most influential proposals are discussed.

7.
Consciousness

Physics seems like such a serious subject that it's surprising to see exaggerated terminology—labeling as the "Grand Unification" of forces what by any definition has to be considerd only partial unification, since it excludes gravity. Hence the escalation of terminology to the "Theory of Everything" once gravity is also encompassed.

But this label—a theory of everything that excludes consciousness— may be a more flagrant exaggeration.

***

There is a minority school among physicists that sees connections between the mind and physics, quantum physics in particular. The majority physicist community views the subject of the human mind and human consciousness as outside the realm of science, and many physicists are openly hostile to attempts to bring consciousness into physics.

We'll be covering later in more detail what some physicists see as the links between physics and consciousness (and we'll also be covering other physicists' adverse peer commentary on these links). In this chapter, we'll review the consensus position among philosophers as to what consciousness is.

What's Physicists' Problem?

Physicists' rejection of including consciousness within the realm of physics may relate to a tendency seen in many fields for the popular understanding of the field to be tied to a classical development of a few centuries back. In algebra, there's often a popular stumbing block at imaginary numbers or noncommutative mathematical systems. In geometry, the non-Euclidean geometries don't always obtain easy acceptance. In art, there's often a public longing for classical portrayals over the more recent surreal or abstract forms (which one's two-year-old can paint). And in philosophy, the most commonly accepted framework for understanding the mind is the dualism set out by Descartes in the seventeenth century: body and mind are two distinct, nonoverlapping realms.

Perhaps it's only by being stuck in Descartes' dualism that physicists exclude consciousness from a theory of everything.

The Modern Philosophers' View

In the almost four centuries since Descartes proposed that mind and body are two distinct, nonoverlapping spheres, philosophers have developed a more modern understanding of the mind as a biological function. The modern philosophers' view of consciousness begins with the fact that the mind is a function of the brain; consciousness is a biological process. Consciousness is not some "other" form of entity; it's part and parcel of what makes us human beings. We all experience consciousness as a biological process, which results (in a complex, not-yet-fully-understood, but nevertheless scientific manner) from our brains and their neuronal and synaptic components.

Digestion is a function of the stomach, breathing is a function of the lungs, walking is a function of the muscles of the legs. Would any scientist, putting out a treatise on our biology, limit the treatise to the organs but exclude their functions?

Why, in the 1997 *Technology Review* listing of "the 14 most compelling questions that fascinate today's scientists and drive their research," would this be the comment on the question "What are the physical origins of memory?":

> The human brain is the most complex object known. While
> an understanding of consciousness still lies more in the realm
> of philosophy than science, we are beginning to address the
> physical nature of information storage in the brain.[27]

Why is an understanding of consciousness so easily dismissed as an insufficiently scientific inquiry, and therefore not included on the list of compelling questions that fascinate today's scientists? And the segue to information storage: why is information storage in the brain being offered—even if apologetically—as any form of substitute for an understanding of consciousness? Could the reductionism possibly be balder?

It is incomprehensible that physicists and other scientists consider excluding the mind from science, especially with quantum physics' long history that has led physicists to ask what their science implies about the nature of reality. Within physicists' own field, they certainly fight this very battle of moving the public out of seventeenth-century constructions of space, time, matter.

But the mainstream view keeps the mind out of physics. Here's Stephen Hawking:

> I totally reject the idea that there is some physical process that corresponds to the reduction of the wave function or that has anything to do with quantum gravity or consciousness. That sounds like magic to me, not science.[28]

Here Hawking is speaking in direct opposition to the work of Roger Penrose, discussed in Chapter 10. Penrose is a prominent advocate of a role in physics for consciousness, and he looks to the reconciliation (quantum gravity) between quantum physics and general relativity to help clarify this role.

At least there's a minority physicists' view advancing a few centuries toward the modern philosophical view of consciousness.

The Competition to the Modern Philosophers' View

The strongest competition today to the general consensus among philosophers is the view of "Strong A. I."—the forceful view of what artificial intelligence can achieve.

Weaker views of what A. I. can do for a philosophical analysis of consciousness tell us that computers and A. I. have some similarities to the brain and the mind, and therefore by studying sophisticated computer technology and architecture we'll learn much about the mind. And Weak A. I. also tells us that our understanding of thinking will aid computer scientists in their efforts developing artificial intelligence. Weak A. I. is acceptable to almost everyone.

But Strong A. I. says that consciousness is nothing more than the aggregation of computer signals or brain waves: there's no distinction between computer processing and thinking, and there's nothing called consciousness that's distinct from a summation of the processes that computers achieve. Strong A. I., though supported by many computer scientists, is rejected by most philosophers.

A possible route to a reconciliation between Strong A. I. and quantum views of consciousness can be found in John Eccles' concept of the psychon as the elementary unit, the smallest granule, of consciousness. We'll be discussing psychons in Chapter 11. If in fact consciousness has elementary units which can be divided no further, then this would lead to a finite limit on how powerful a computer would need to be in order to match the mind's capabilities. A computer would fully replicate the mind as soon as it could process the quanta of consciousness as fast as the brain can. So Strong A. I. becomes equivalent to consciousness at the one-psychon level. A computer achieves consciousness when its processing is at the one-psychon level. Perhaps psychons are the force-carrying particle for the force of consciousness.

But generally, those physicists finding a role for quantum physics in understanding consciousness also reject Strong A. I. This is a main point of Roger Penrose's *The Emperor's New Mind: Concerning Computers, Minds, and the Laws of Physics*, which we'll be discussing in Chapter 10.

***

First, in the interest of full disclosure, let's explore some clear misapplications of quantum physics.

For those with no inclination to permit quantum physics to enter the world of the "soft sciences," these social-science misapplications of quantum physics may reinforce your views. But perhaps they'll have the opposite effect; perhaps you'll distinguish these extremes of "junk science" from some more plausible applications.

8.
Early Extensions of Quantum Physics into the Social Sphere

Quantum physics—with its uncertainties, and with the role of the observer being brought into the science of physics—has been a rich field on which to project stances from the social sciences. When politicians have tried to make political points by drawing political conclusions from quantum physics, their motives are entirely transparent and they can be seen to be taking the most superficial of understandings and twisting them for their own purposes. Nor have theorists—even physicists—been received particularly well when they've drawn from quantum physics to suggest conclusions about religious beliefs, or emotions, or consciousness.

***

Politicians Co-Opt Quantum Physics

Physicists and science historians have noted attempts at political applications of quantum physics since the earliest years of quantum physics' development.

Science historian Helge Krage, for example, devotes a full chapter to "Physics and the New Dictatorships" within his extensive *Quantum Generations: History of Physics in the Twentieth Century.*

During the Nazi era in Germany, Kragh notes, Aryan physics rejected relativity theory and other developments of modern physics. This rejection derived in part from the fact of the Jewish background of Einstein and some of the other developers of these theories. But it also derived from the sense that Aryan culture is more based on concrete, observable nature, not the abstract formulations of mathematics-dependent modern theories.

The early Soviets also debated how quantum physics applies to their political philosophy, in their case debating whether quantum physics was consistent with Marxist-Leninist thought. Again drawing from Kragh's "Physics and the New Dictatorships," we see one strand of Soviet thinking finding the new physics to be idealistic or subjective, and therefore contrary to correct Marxist principles of objectivity and realism. But we also see another strand arguing that the new physics is "a brilliant confirmation and enrichment of dialectical materialism."[29] Presumably, Marx's thesis/antithesis dialectic is echoed by the uncertainty principle's tradeoffs and conflicts, and by quantum particle/wave duality.

Of course, both the Nazi and Soviet political spins on quantum physics were serving other purposes besides looking to physics to confirm what is politically correct. Military ambitions and political power grabs were involved. But quantum physics, as well as relativity, can be lightning rods for the social sciences. Quantum physics and relativity question the contemporary understanding of reality, and therefore can be used for or against various political points of view in times of political upheaval.

Kragh also traces the political uses of physics through the last decades of the twentieth century, where one strand of radical political thought looked to challenge physics—and contemporary science in general—as constructions of masculine thinking. This strand also challenged the scientific method and much that today's science is based on.

And Stephen Hawking, in a 1980 lecture, noted that China, under the Gang of Four, had official dogma regarding another specific point of physics: whether we'd ever stop finding yet smaller elementary particles than what had been found to date. The dogma, purportedly deriving from politically correct Maoist thought, was that this will be an infinite quest.

But more on point to this book is when it has been not politicians but physicists, also during the last decades of the twentieth century, who have made attempts to extend physics into the social sciences.

Physicists Co-Opt Quantum Physics

Physicist Fritjof Capra's *The Tao of Physics* presents a different late-twentieth century perspective than the perspective that rejects modern science. We've mentioned his work earlier, as one of physics' perspectives on the nature of quantum reality. Capra discusses the perspective that physics is mysticism. Rather than understanding physics as based on rigid "masculine" thinking, *The Tao of Physics* presents physics as consistent with ancient Eastern mysticism. Kragh notes mainstream physicists' view that Capra relied on outdated physics, and Kragh's evaluation is that the thinking of the adherents of Capra's approach "made no impact at all on mainstream physics."[30]

But Capra was not the first quantum physicist to suggest that quantum physics has socio-psychological implications.

Niels Bohr, an early-twentieth-century founder of quantum physics, extended the wave/particle dualism (which he and his predominantly Copenhagen colleagues incorporate within the complementarity principle) to psycholgy, biology, and cultural questions in general. Kragh notes Bohr's pairing of emotions with our perceptions of emotions in this same framework of complementarity, standing "in a complementary relationship analogous to situations of measurement in atomic physics."[31]

Einstein opposed Bohr's views on this, and "described the Copenhagen interpretation sarcastically as 'the Heisenberg-Bohr tranquilizing philosophy—or religion?' And he added that 'it provides a gentle pillow for the true believer from which he cannot very easily be aroused.'"[32]

More recently, Roger Penrose, in his 1989 book *The Emperor's New Mind: Concerning Computers, Minds, and the Laws of Physics*, proposes very specifically that the human mind is to be understood through the resolution of the conflict between general relativity and quantum physics. And remember: this is the same Penrose whose "report card" on physics' theories falls on the cautious side of mainstream.

***

Penrose's thesis generated widespread, mostly disapproving reaction. Within a year of the publication of *The Emperor's New Mind*, thirty-seven scientists published their criticisms in a special *Behavioral and Brain Sciences* section.[33]

But Penrose's thesis is intertwined with the thesis of this book. So this—and a discussion of what so bothered thirty-seven physicists—certainly deserve their own chapter.

9.
Doris Lessing

A bit of a digression here. But in her own field as a great writer, Doris Lessing has gone through what may be an analogous experience—and an analogous set of reactions—to Roger Penrose's experience with his theories of the new mind of the emperor.

***

The Long-Delayed Nobel Prize

Doris Lessing has had a long and distinguished writing career. Her name has appeared for years on critics' short lists of possible winners of the upcoming Nobel Prize for literature.

Lessing's varied writings for many years were fully in the realistic mode. Based in part on her own life in southern Africa and later in London, her writings are striking for the detail and depth of her understanding of people—how we relate, what drives us, what we're thinking and feeling.

Beginning in the 1950s, Lessing published a five-volume "masterwork" called the Children of Violence series. This series took Martha Quest, its main character (the outlines of whose life reflect Lessing's) from her young adulthood in southern Africa through her migration to London.

The first four volumes of this series, and the first part of the fifth, were in style similar to Lessing's great earlier work, *The Golden Notebook*. The reader knows there are kernels everywhere of Lessing's own life, but what is mostly happening for the reader is a visit to the bodies and lives and minds of the books' characters.

The reader lives the experiences of Lessing's characters. The reader's real life intertwines with the characters' real lives.

Partway through Volume 5, which is called *The Four-Gated City*, Lessing leaves the realistic mode.

I'm not sure how else to characterize what happens.

A nuclear explosion has taken place, and Martha Quest's world has suddenly lost all of its modern conveniences. There is much destruction, and there is no longer any electricity or other utilities. The modern world has ceased to function the way we know it.

Schooling no longer seems relevant, and children seem to have little interest in reading and writing. But they are very interested in new forms of communication, which seem to involve telepathy.

And a world of higher consciousness emerges, new psychological modes and new social interactions.

Well, this was all too much for many of Lessing's fans. They did not buy into this nonrealism. It was not what they learned to love Lessing for.

Reviewers made a specific point of noting that there was a fully realistic Lessing for more than four volumes of the Children of Violence series, and that there is a specific break from realism to nonrealism at a specific sentence on a specific page in *The Four-Gated City*.

It seemed important to many that a black-and-white distinction was to be made. For years Lessing had been completely realistic. Her characters, even though fictional, had thoughts and social interactions that could really happen. Characters spoke out loud, and others heard with their ears and responded with their mouths. In these early works, even the "reading" of nonverbal cues is fully realistic—perhaps you could say hyperrealistic, Lessing's description of the interactions are conveyed so clearly and in such depth. But never, until black became white, until telepathy and paranormal phenomena, did Lessing leave the realistic mode.

The sense of betrayal and abandonment was palpable. Lessing was intensely realistic up through part of Volume 5. Then at a specific sentence, and beyond, she left the realistic mode.

Lessing was not to be deterred. Although she followed *The Four-Gated City* with other books returning to the realistic mode, she also made a brief detour back into the mystical with *Memoirs of a Survivor*. There, sitting in her room, the protagonist can look through walls and see, hear, even participate in another world, through the solid wall of her apartment, into another place and time.

Then Lessing went off the deep end entirely—detractors felt—with her next five-volume series, Canopus in Argos: Archives.

This Canopus cycle of novels conveys grand intergalactic struggles among the universe's great civilizations. A benign colonizing civilization sees potential for the beautiful newly developing planet Rohanda (that's us), and ambassadors from this colonizing civilization have borne themselves into bodies of Rohandans to supplement their guiding along of our development.

There's much more, of course (the universe's forces become misaligned, and Rohanda becomes Skikasta, the broken planet). But for this book, it's the reactions of Lessing's great fans that I find most interesting.

Many of them are not enthusiatic. Among these, the general thrust of dissatisfaction centers around disappointment that Lessing is wasting her talents on this nonrealism of space fiction.

What's particularly clear in the minds of many is that there is a real version of the world, and there is a nonreal version of the world. We all know exactly when Lessing has crossed into the forbidden land, and at that point many of Lessing's readers, including many of her most devoted fans, reject her efforts.

Why is this? Why is it so clear to Lessing's critics (even the ones who love her real-mode works) that she's exited the real world at very specific points? Why is it that any fictional situation is allowed, even if implausible, as long as it uses only the vocabulary and structure of our corporeal world? But it's a different world entirely—a rejected, nonexistent world for many—the moment Lessing crosses the threshhold into nonrealism.

Roger Penrose can ask a similar question.

10.
The Emperor's New Mind

Roger Penrose is a prominent University of Oxford mathematician who has made major contributions to modern physics. His 1989 book—aimed at the popular market although with significant scientific substance—speculated on the nature of consciousness.

***

We discussed in Chapter 2 the path of twentieth-century physics, and the rankings of centuries of scientific thought that Roger Penrose presents in his 1989 book, *The Emperor's New Mind: Concerning Computers, Minds, and the Laws of Physics*. Penrose mentions that he's been asked how he would rank a theory of physics—twistor theory—that he himself has been developing over the years as a proposed reconciliation of quantum physics with general relativity. Penrose answers that twistor theory can be placed no higher than "tentative."[34] It certainly can't be any higher up on the scale that Penrose would rate his speculations on consciousness, with which he closes *The Emperor's New Mind*.[35]

Penrose is clearly of the school that consciousness goes beyond the simple accumulation of more and more complex algorithmic capabilities. Therefore, Penrose is not within the Strong A. I. school of artificial intelligence that advocates that computers either now have a mind, or at least will soon have a mind once we have exceeded a critical mass of computational capacity and speed.

Penrose supports his argument against the Strong A. I. philosophy by invoking the Gödel incompleteness theorem, by which the German mathematician Kurt Gödel proved that no mathematical system—no formal system of logic of any type—can ever be truly complete, in the sense of proving everything within its scope. This, in Penrose's argument, contradicts any claim that a mechanical computational system will ever replicate the complexity of the mind and human intelligence.

Penrose speculates that consciousness involves access to the universe's idealized concepts; these are the Platonic ideals of centuries-old philosophy. In Plato's formulation, it is not our typically understood physical world that is real; what's truly real are forms and ideas. The physical world is a mere shadow of the real world of forms and ideas.

When the mind perceives one of the mathematical concepts of Plato's worldview, the mind is making contact with this world. Our experience of grasping a concept is a holistic experiences of seeing at once, as a whole, the solution to a problem. Or, as Penrose cites, it's Mozart discussing how he seizes at a glance an entire musical composition: "It does not come to me successively . . . but in its entirety that my imagination lets me hear it."[36]

Before our mind reaches these kernels of understanding, Penrose proposes, a physiological process within the brain allows the brain to form these ideas. The process involves physical brain activity—rapid trials of combinations of growing and contracting dendritic spines, which stretch out to the synapses that separate a nerve cell from its neighbor.

These trials take place under the radar screen. They must be short-lived, because the nonvalid trials would otherwise be detected through the electromagnetic fields that they would produce. And the trials must take place below the one-graviton level.

Now . . . the one-graviton level . . . this is exciting stuff! Remember: the graviton is the particle that, according to quantum physics, transmits the force of gravity. The graviton is indivisible—it's an elementary particle, and therefore it gives us the lower limit for the size of a granule of gravity. The smallest granule of gravity would be that transmitted by one graviton. And since gravity reshapes space, another way of saying the same thing is: the smallest disturbance of the shape of space that can be produced is that produced by one graviton. Above this level, we are operating in the world of measurable certainty. At this level and below, we are operating in quantum physics' world of uncertainty, a world where things don't exist with single-point definiteness, but instead have various probabilities as to the form in which they can precisely exist.

How small is this? Penrose makes a rough estimate that, measured in terms of mass, the one-graviton level is one ten-millionth of a gram, 10^{-7} grams, which for the quantum world is very big. A hydrogen atom has mass one hundred million billion times smaller (that is, mass of 10^{-24} grams).[37]

Gravity is such a weak force that it requires mass enormously larger than an atom before gravity's elementary particle can transmit or sense gravity. But we'll sense a hydrogen atom electromagnetically long before we'll sense it gravitationally. Penrose is proposing a quantum gravitational window, without detection aided by other forces, and he's relying on the additional constraint of a short time duration to avoid electromagnetic detection.

So our brain has a window of opportunity within which to toy with possibilities for dendritic spine construction. How does the brain settle on its ultimate choice?

Penrose goes on. In part, the construction is influenced by the physiology and chemistry of its environment. So the construction depends in part on our emotional state and on the preexisting state of our brain and its connections.

But what provides the core decision-making criterion? How is a final dendrite construction settled on when our mind grasps a concept or glimpses a new symphonic work?

Here, Penrose takes this even further. His answer is quantum gravity, which is also the (still not found) answer to the question of how general relativity is to be reconciled with quantum physics.

Penrose has frequently collaborated with Stuart Hameroff, who has extensively studied microtubules, which give shape to our neurons and through which neuronal chemicals pass. Hameroff, tracing the evolution of life, marks the incorporation of microtubules into the modern cell as taking place about 1.5 billion years ago, as part of a general symbiotic merger of previously independent organelles (cell parts). A billion years later, during the Cambrian period which began 540 million years ago, there was a vast and abrupt emergence of varied lifeforms—the Cambrian explosion—which Hameroff attributes directly to the early precedents of consciousness that microtubules permit.[38]

Penrose and Hameroff propose these microtubules as our brain's link—through *orchestrated reduction*—to the collapse (reduction) of the quantum wave function: many neuronal microtubules, acting in concert (orchestrated), create an act of consciousness linked to the quantum physical world.

Penrose is not presenting the full story; he is looking first for the correct understanding of quantum gravity to be developed. It is Penrose's belief that, through this understanding, the phenomenon of consciousness may be elucidated.

Penrose emphasizes the noncomputable nature of consciousness, and he expects to find an analogous noncomputability in our ultimate understanding of quantum gravity. But he warns: consciousness "will fit only very uncomfortably into our present conventional space-time descriptions."[39]

Reactions to *The Emperor's New Mind*

In 1990, the year after the publication of *The Emperor's New Mind,* the journal *Behavioral and Brain Sciences* published a summary by Penrose of *The Emperor's New Mind,* as well as peer commentary on the book. Commentary from thirty-seven scientists was published, with most of this commentary raising questions about various aspects of Penrose's conclusions. The journal also included Penrose's response to the peer comments.[40]

The criticisms concerned many aspects of Penrose's mathematical, philosophical, and biological arguments, as well as his applications of physics to the realm of consciousness. Penrose has not backed away from his arguments, either in his *Behavioral and Brain Sciences* response, or in his expansion of his theories in his subsequent book, *Shadows of the Mind: A Search for the Missing Science of Consciousness,* and elsewhere.

***

The idea that quantum physics plays a role in consciousness remains a minority view among physicists. But not a minority of one.

11.
More Physicists' Thoughts About the Mind

A number of other physicists besides Roger Penrose have published research and discussion about the physical basis of consciousness and the mind. As is true for Penrose's work, these other physicists' ideas are not universally accepted, but I present them as they are proposed.

***

Atomic Consciousness

Physicist Fred Alan Wolf published in 1981 (updated in 1989) his book *Taking the Quantum Leap: The New Physics for Nonscientists*. Based on his own and others' research, Wolf presents the concept of human beings having become "atomically conscious"—having gained the capabilty of using our brains to operate within the world of quantum physics. The idea is that this capability has been an evolutionary advancement of human beings, dating back only three thousand years in our evolution. Atomic consciousness is a link in the chain between human consciousness and quantum physics.[41]

Wolf proposes atomic consciousness as a mechanism for the creation of humans' modern single mind, which broke through the barrier of separation between the two hemispheres of prehistoric humans' bicameral mind. Here Wolf is referring to Julian Jaynes' radical 1976 book, *The Origin of Consciousness in the Breakdown of the Bicameral Mind*. Jaynes proposes that ancient peoples, earlier than three thousand years ago, were not conscious in today's sense, did not have a sense of self, a sense of "I." These ancient beings did not clearly distinguish the inner self, the world of their minds, from their external environment. But the precursor of modern consciousness was presented to these ancient beings as auditory hallucinations, originating in one of the brain's hemispheres during novel or stressful situations. These auditory hallucinations are reminiscent of tales from mythology and from the Bible, and they have relics today in phenomena of schizophrenia, hypnotism, and even religion and poetry. Jaynes focuses on the eventual social learning by which the understanding of these hallucinations transformed to modern consciousness, and Wolf focuses on the biology and physics of this transformation's mechanisms.

More specifically, relying in large part on work of the Australian mathematician L. Bass,[42] Wolf discusses in detail a physical, concrete connection between our brains and quantum physics' indeterminate world of probabilistic potentialities. In particular, Wolf identifies a specific pair of hydrogen atoms that connect the brain to the quantum world.

These two hydrogen atoms are part of a molecule that is at the tail end of an enzyme that operates a protein "gate" that opens and closes channels that connect a nerve cell to its neighbor.

Like all atoms, these two hydrogen atoms operate within the quantum world. This means that most of their existence is only as potential atoms in possible locations. When they're observed, however, their existence is momentarily stabilized at a specific location.

These atoms have two possible locations in which their existence may stabilize: they can stabilize in a location that opens the protein gate, or they can stabilize in a location that closes the protein gate. If they are not observed, they do not stabilize. But alternatively, they may, through the mechanism of human atomic consciousness, be observed located at the site where they open the gate that allows the nerve cell to fire towards its neighbor. Or they may be observed located at the site where they close the gate and prevent the nerve cell from firing toward its neighbor. Either way, within our nerve cells we have developed devices that can choose to observe the enzyme's activity at the atomic level, thereby triggering its quantum stabilization. Ultimately, our brain senses the aggregation of these and other atom-level quantum observations, creating an atomic consciousness.

Dendrons and Psychons

Quantum effects seem so small that it is natural to question whether it is conceivable that these effects can be experienced at the level of human biology. Fred Alan Wolf goes back to the basic principle—Heisenberg's uncertainty principle—to show that this is conceivable: in a 1989 *Journal of Theoretical Biology* article, Wolf demonstrates this not for his own particular proposed biological mechanism, but rather for one proposed by John Eccles, who shared the 1963 Nobel Prize for medicine for his work on ion processes within nerve cell membranes.

Eccles's proposed quantum-conscious events take place at a vesicle (thin-walled sac) of our nervous system's synapses. Wolf estimates the mass of this vesicle and the size and speed of its movement. With the mass known, the uncertainty principle specifies a very small minimum for the combined uncertainty in our measurement of speed and position. Wolf's calculations show that these magnitudes are small enough to subject these synaptic emissions to the realm of quantum physics,[43] confirming mathematically the conclusion drawn in Eccles's 1986 article "Do mental events cause neural events analogously to the probability fields of quantum mechanics?"

Eccles is driven to a large extent by the need to respond to objections raised to his work that mind-to-brain action violates physics' law of conservation of energy: If our physical brain constitutes a closed world of energy and matter, how can the nonphysical world of the mind affect this closed world of the brain without violating the law of conservation of energy? Won't there be bound to be a net change in the physical system's energy if it is influenced by the world of the mind?

To answer this question, Eccles first notes that vesicular emissions that result in neural firings are themselves chance quantum events subject to quantum probability. Eccles gives examples of probabilities anywhere in the range from 5% to 90% for a triggered vesicle actually firing. Eccles' main point then becomes that what the mind does is alter the probability of these emissions being triggered, and this is how he gets around any introduction of mind energy into the physical brain, thus avoiding a violation of the law of conservation of energy. The role of the mind (mental events) is merely to alter the probabilities of emission from vesicles that already had a probabalistic spectrum of disposition for firing.

So the mental event for Eccles involves the selection of the vesicles that are the correct vesicles to influence in order for the brain to perform the intended physical event. This selection is accomplished according to a "learned inventory." Eccles admits that this "may seem like a clumsy method" by which the mind operates, but he finds this to be a necessarily complex strategy for the mind to be able to address the enormous number of possible activities of the brain.[44]

Eccles, in later work,[45] again makes the point that the mind (the mental event) does not take an initiative in bringing on the synaptic emission (the "exocytosis in the presynaptic vesicular grid," in Eccles' terminology); rather, the extent of the mental event is only to select vesicles—which already have a probabalistic profile of likelihood of firing—and influence this probability of firing in the direction of the mind's intention.

Eccles develops this hypothesis more fully, noting that a single mind-to-brain quantum phenomenon is unlikely to be large enough in magnitude to create a meaningful physiological (brain) phenomenon. So there must be concerted effects in both the functionality of the mind (which selects vesicles for which it will influence the quantum probabilities otherwise in place) and the functionality of the brain (which requires concerted messaging from multiple sources before it can take a significant action). Therefore, Eccles looks not to single vesicles, but instead to thousands of vesicles attached not only to individual neural dendrites, but to a grouping (*dendron*) of these dendrites. This is the concerted brain effect. And each dendron is penetrated by a *psychon* (mental unit, manifested by a probability field) through which mental intention acts with concerted effect.[46]

Computer scientist Ying Liu presented a detailed theory of this at a 1995 conference of the International Society for Optical Engineering. Presentation of this subject was largely motivated by investigation of possibilities for artificial intelligence. Liu models the psychon to reflect the nature and degree of feelings (emotions), and to reflect the historical states of the psychons, mind, and brain. Psychons are probability fields—not physical particles—and they provide a mathematical structure in which the dendrons operate: they specify for the physical brain the probabalistic expectations of the mind.[47]

Both Eccles and Liu also discuss mechanisms and models for the brain-to-mind direction of interaction. That is, the brain's activities also are registered in the psychons and consequently affect our mental state. Thus, Eccles' "microsites"—the specific location of these quantum neural events—are proposed to connect our consciousness with our biology.

Eccles' work has received its share of skepticism and adverse commentary. Kitcher, for example, considers that Eccles is only "waving dimly in the direction of quantum mechanics," finding Eccles' proposals insufficiently detailed and supported. Kitcher concludes that Eccles has "descend[ed] into dogmatic mysticism."[48]

Family of Quantum Switches

University of Cambridge physicist M. J. Donald proposes that our brains have several different quantum switches, beyond those proposed by Penrose, Wolf, and Eccles, and he focuses in particular on sodium channel proteins, which are involved in the brain's neural firings. Donald proposes a model of the brain as a family of quantum switches.[49]

Endorphins

Richard A. Mould, physicist at State University of New York, Stony Brook, introduces the term "common mechanism" to define biological entities involved both in consciousness and in quantum physics' collapse of the wave function. He adds to our proposals for common mechanisms the class of proteins called peptides, of which endorphins are an example. The endorphin proteins are involved with eliminating pain and with creating feelings of euphoria, and Mould demonstrates that their operation falls within the range of quantum effects subject to the Heisenberg uncertainty principle.

Mould hypothesizes that, within the framework of our evolutionary advancement, it has been advantageous for us to develop a mechanism—an inside observer process—by which our psychological and biological states evolve in parallel. We become more skilled at deploying these inside observer processes, both as we develop as individuals throughout our lifetimes, and as we have evolved as a species. For example, early in our evolution, as we were just gaining consciousness, pain and pleasure were our first conscious experiences, and our conscious states developed in concert with changing patterns of quantum probability amplitudes that had the effect of decreasing pain and increasing pleasure.[50]

Deterministic and Nondeterministic Quantum Phenomena

Berkeley physicist Henry P. Stapp discusses yet another "common mechanism"—the calcium ions in neurons—as biological entities that operate in both the world of consciousness and the world of quantum physics' collapse of the wave function.

Stapp is of the school that quantum physics is at its heart a theory of the interaction between mind and matter. Quantum electrodynamics (QED), the well-established quantum theory of electromagnetism, is the aspect of quantum physics that operationalizes the quantum physics of the brain, since the brain operates biologically through physics' electromagnetic forces. Stapp makes a point of noting that, whatever high-energy theory comes to be understood as a quantum theory of everything, this theory will reduce to QED in the energy range of atomic and molecular interaction.

Stapp applies a bit of a magnifying glass to the phenomenon of quantum consciousness, breaking the phenomenon into three processes that shed light on how consciousness interfaces with the world of quantum physics.

Quantum physics has both important deterministic elements and important nondeterministic elements. One of the three processes involved in the physics of the world is the highly reliable and accurate set of deterministic phenomena of quantum physics, which control how the world works, subject to the Schrödinger equation for the continuous causal time evolution of the quantum wave function. This process proceeds deterministically, neither subject to the chance effects of the laws of probability nor reliant on effects of consciousness. But this first process is certainly mysterious enough: many different quantum possibilities are, in parallel, systematically rolling forward in time. Their nonweird aspect is that each possibility has a predictable rollforward step by step. But their weird aspect is that, unobserved, many distinct possibilities are all rolling forward separately, in parallel, along different paths.

The second of the three processes is the mind's—or the observer's, or the experimenter's—decision as to where to focus attention. No wave functions collapse unless they are attended to, and the universe is filled with uncountable possibilities for focus of attention.

What do you want to look at? Consider? Focus on? Attend to?

This is a choice made by the mind.

The third process is quantum physics' nondeterministic phenomenon—the collapse of the wave function, subject to the probabilities of quantum physics' statistical rules—but not determined with certainty (only with likelihood; the most likely possibilitity does not always occur). Amazingly, the same Schrödinger equation that governs the first, deterministic process has hidden within it the probabilities for each possible collapse. One set of arithmetic rules applied to the Schrödinger equation gives us the deterministic but parallel roll-forward of unobserved possibilities. Another set of arithmetic rules applied to the same equation gives us the probabilities for how the quantum wave function might collapse upon observation.

After the mind selects the focus of attention, and through its common mechanisms observes and is affected by the outcome at which the quantum wave collapses, the physics is handed back to deterministic unfolding in accordance with the Schrödinger equation for the quantum wave function. The operational environment of the physical brain and (if you subscribe to the views of this chapter's physicists) the conscious mind are all subject to the world of quantum physics.[51]

Some of Stapp's commentary on his theory also applies to theories we will return to many chapters from now: "Thus the whole range of science, from atomic physics to mind-brain dynamics, is brought together in a single rationally coherent theory of an evolving cosmos that consists of a physical reality that represents information, interacting via the laws of atomic physics with the closely related, but differently constituted, psychical aspects of nature."[52]

Stapp can't beat Penrose's record of the near-immediate publication of criticism of Penrose's work by thirty-seven fellow scientists. But Stapp comes close. Eighteen objections to Stapp's work were all published by one commentator, Ulrich Mohrhoff, as "The world according to quantum mechanics (or the 18 errors of Henry P. Stapp)." Stapp published clarifying responses, standing by his approach to keeping the mind within our understanding of physics and nature.[53]

Criticism from the Other Side

None of the physicists we've discussed in this chapter hypothesize that the mind is in direct control of the physical processes of the brain's operation. The mind is not directly willing the closing of gates, the firing of neurons, or the release of chemicals.

All of the physicists we've discussed in this chapter hypothesize that there is active conscious involvement in the mind's processes as quantum physics' observer. These physicists provide various possibilities for mechanisms that connect quantum events with the mind and brain.

There are shades of difference in these physicists' views of how the mind affects the brain. For example, Wolf has the mind selecting observations that will confirm collapse at the desired result, whereas Eccles has the mind's psychons actually influencing the likelihood of collapse at the desired result.

But all of these physicists go beyond what is the majority view among physicists. All of these physicists believe that consciousness is a proper field for study within physics, that consciousness affects the physical world, and that consciousness is a phenomenon of the physical world.

Ironically, the views of Penrose and of this chapter's physicists are criticized from the other side also. These views don't go far enough for the most comprehesive theorists of new physics and the mind. For these theorists, whom we'll be meeting in this book's Part Four, it is ultimately reductionist to focus on locating common mechanisms at local corporeal sites. For some, it is wrong to "privilege the collapse of the wave function."[54]

Many of Part Four's new physicists of the mind incorporate within their theories an accumulation of new physics phenomena that seem outside of the well-worn path towards quantum gravity. These are the phenomena that we'll be reviewing in Part Three, New Physics. An understanding of quantum gravity is an important aspect of this new physics, but so are a number of additional phenomena.

Old Physics and the Mind Becomes New Physics and the Mind

New physics has many connotations, some dating back to the earliest days of modern physics. For this book, I'll use the term new physics to indicate two characteristics of scientific inquiry.

For new physics, the end of physics will not occur at quantum gravity, when quantum physics is reconciled with general relativity. A funny thing happened on the road to quantum gravity: a number of additional physical phenomena have emerged off-road, not on the path toward quantum gravity as the end of physics. Perhaps most notably, these are phenomena of entanglement, phenomena of superluminal (faster than the speed of light) linking of quantum events and particles. We take all of Part Three of this book to review the road to quantum gravity as well as various other phenomena of new physics.

The second characteristic of new physics is that it looks beyond reductionism or atomism. New physics incorporates physics of holism, of emergence, of coherence. Dissection has taken old physics very far, but new physics asks if it is time to look for a whole greater than the sum of its parts. Marshall and Zohar, for example, cite the underlying quantum vacuum as a phenomenon of quantum relational holism: modern quantum field theory "describes all existence as an excitation of the underlying quantum vacuum, as though all existing things were like ripples on a universal pond." And they wonder if consciousness is an example of an emergent phenomenon—not explainable by properties of its components, and perhaps needing quantum processes to explain.[55] Emergent processes are self-organizing: they reflect "the tendency for things to arrange themselves into new patterns that, as organized wholes, possess new types of structure and new qualities [or] properties."[56]

So, radical as they are, the quantum switches and endorphins and calcium ions and psychons and atomic consciousness and other common mechanisms that we've discussed so far in this chapter do not go as far as the radical theories of new physics we'll be discussing in this book's Part Four.

Penrose, I believe, bridges the gap between old physics and the mind and new physics and the mind. His microtubules strike me as a reductionist mechanism which looks for a site, a location, where the mind/brain connection resides. And clearly Penrose is driven by the quest to reconcile quantum physics with general relativity, since he ties solving this mystery to solving the mystery of consciousness. But on the other hand, the Platonic ideals that Penrose's microtubules reach out towards, and the noncomputable nature of consciousness—these are phenomena of the whole, phenomena we'll see again in some of Part Four's radical theories of new physics and the mind.

Wolf, too, theorizes about the matching of events taking place in the physical world to events in internal space, mental space. This concept is also central to several of Part Four's radical theories of new physics, including Radical Theory Number One, Pitkänen's topological geometrodynamics, which connects mathematics' only two complete number fields, one field mapping the physical world and the other field mapping the mental world. Wolf discusses quantum psychodynamics— the quantum physics of psychology—including psychophysical parallel planes and the mathematics of the transformation of feelings into thoughts, and the transformation of intuitions into sensations, and back again. Much of Wolf's physical/mental mapping presages later work of new physics and the mind.[57]

The Implicate Order

Although the discussion of radical theories of new physics and the mind will be postponed until after, in Part Three, we learn more about phenomena of new physics, we'll close out Part Two with an introduction to what may be the beginnings of new physics and the mind: physicist David Bohm's implicate order.

A number of physicists who are considering the physics of consciousness have been influenced by Bohm's work, including his 1980 book *Wholeness and the Implicate Order*. This is a work drawing far-reaching conclusions about both physics and consciousness, from a starting point of the concept of wholeness, and Western and Eastern insights into this concept.

Bohm's earlier publications included work on quantum theory and on relativity. He developed the hidden variables variation on the Copenhagen Interpretation of quantum physics, which diverges from the standard Copenhagen Interpretation stance that realites do not exist until they are observed. In the hidden variables interpretation, there is an underlying set of unknown and unseen information that governs the apparently indeterminate quantum outcome, implying that reality is well-defined even in the absence of observation.

The hidden variables approach was slow to gain widespread acceptance, in large part because it implied nonlocality—instantaneous communication of information. For hidden variables to simultaneously influence distant phenomena, it is argued, information must be communicated faster than the speed of light, which seems to contradict physics' prohibition against anything, even information, traveling faster than the speed of light.

But within the last decades of the twentieth century, superluminal phenomena—with entanglement being an example—have in fact been experimentally confirmed and have become a major focus of new physics inquiry as we enter the twenty-first century.

Bohm continued the development of his theories and proposed that the classical world—the Explicate Order of fixed spacetime grids in which matter has specific position and boundaries—is a shadow of a deeper reality, the Implicate Order. The Implicate Order is a nonlocal whole, within which processes emerge that create, in the Explicate Order, appearances of temporary bounded localized particles, before these forms unfold back again into the Implicate Order. Our scale—our large size and slow speed of motion, compared to the quantum world—helps create from the underlying Implicate Order the appearance of independent localized objects. It is within the Implicate Order that Bohm finds the deeper nature of human consciousness.[58]

Bose-Einstein Condensates and the Sense of "I"

As a preview of Parts Three and Four, we end Part Two with a theory of new physics and the mind that reflects applications of Bohm's work.

Entanglement is a phenomenon of new physics that we will be discussing in Part Three, as is the Bose-Einstein condensate, another now experimentally verified nonlocal application of quantum physics.

Danah Zohar looks to explain the unity of consciousness—the single sense of self, or "I," that we each feel—through a quantum theory of consciousness that relates the holism of "I" to an analogous holism of quantum physics. A central mechanism for Zohar's approach is the Bose-Einstein condensate, which can be formed as a result of supercooling, to temperatures just above absolute zero. Such a condensate consists of particles that have identical quantum states, and as a result this coherence—this exact alignment of particles—permits the aggregate condensate to take on frictionless characteristics (superfluidity), which can also result in superconductivity. This phenomenon of coherence is also exhibited by the alignment of light to create lasers' power.[59]

In Zohar's model, chemicals within our blood cause alignment of electromagnetic force particles within the molecules of our brains' neurons. A unity of consciousness is created when molecular dipoles are so well aligned as to create the fully coherent pattern of a Bose-Einstein condensate, which constitutes a unified field across the brain and creates a uniform ground state of consciousness. Our thoughts, our perceptions, our emotions are ripples on the condensate. These temporary excited states are etched onto the character of the condensate, creating, sustaining, changing our emergent quantum self.[60]

Ian Marshall, Zohar's frequent collaborator, notes that this emergence, this quantum relational holism, is neither a dualist sense of mind nor a reductionist sense. Unlike the dualist perspective, mind and consciousness are part of the natural order, part of science, not something beyond the real, something divine, inexplicable. And unlike the reductionist perspective, an explanation of consciousness cannot be reduced to its physical characteristics, its spatial and temporal extent and its mass.[61]

***

As Zohar commented in 1990: "Consciousness is a fact . . . A philosophy or a science that can't account for consciousness is a necessarily incomplete philosophy or science."[62]

Zohar continues this thread in the introductory paragraph to a 1995 *Minds and Machines* article: "A physical model of consciousness . . . is what we need, a model which can place human beings and the whole rich world of our conscious life within the wider context of physical reality. It is only through such a model that we can see ourselves as fully part of the cosmos."[63]

We have more physics to establish before we return to the mind. Let's start with quantum gravity.

Part Three
New Physics

Quantum gravity is the jewel in the crown of the Theory of Everything. It is where the extended lines of relativity and quantum physics eventually intersect.

But the royal wardrobe includes more than the crown, and there seem to be precious stones throughout the royal robe. New conceptual focuses—extra dimensions, entanglement, tunneling, dark matter, dark energy, black holes, entropy, information, condensed matter, the history and future of the universe, the role of mathematics, even the role of art—are jewelry in their own rights and seem outside of the straight-line paths toward quantum gravity.

Kafatos and Nadeau, whom we'll meet again in Part Four's countdown of radical theories of physics and the mind, date the emergence of new physics all the way back to the beginning of the twentieth century. They note the late-nineteenth-century encouragement of physics students to migrate to other fields, since—as physicist Lord Kelvin noted—"only two small clouds" in physics remained to be cleared up. Kelvin was not able to foresee the enormous implications of solving these two mysteries—Michelson and Morley's failure to detect the theoretical ether through which light travels, and the bunched frequencies at which a simple black body radiates. In fact, the solution to these two mysteries led to both major strands of the early twentieth century's "new physics"—Michelson-Morley to relativity, and black-body radiation to quantum physics.[64, 65]

The components of Part Three's new physics are late-twentieth-century challenges to the standard models of physics developed earlier in the century. In Part Three of *New Physics and the Mind*, we'll look at new physics' trees, in order that in the book's final part we'll be educated rangers for the forest of new physics.

Part Three's arts & entertainment section is: arts & entertainment. More specifically: What is the role of art in the new physics? Of literature? Of religion?

***

Part Three Summary

Fifteen contemporary areas of physicists' inquiry are discussed. Overriding these areas of inquiry are questions of whether they stretch the twentieth century's standard models of physics beyond their breaking points.

Chapter 12. Quantum Gravity. Quantum gravity is the Holy Grail of modern physics. It reconciles the two major strands of twentieth-century physics—relativity, which is largely a theory of gravity, and quantum physics. Modern theories explaining quantum gravity range from essentially linear extensions of existing science to radical reconfigurations of the nature of the physical world.

Chapter 13. Extra Dimensions. String theory, as well as a number of other proposed theories of modern physics, suggest that hidden dimensions influence our perceived physical world. Some proposals assume that these extra dimensions are hidden because they are small, but other proposals consider large and even infinitely extended extra dimensions. Experimental physicists consider how we can observe these extra dimensions.

Chapter 14. The Universe. We keep getting surprised by new phenomena of the universe, such as dark matter and dark energy, which observation tells us must exist but which theory can't yet convincingly explain. Some explanations propose extravagant new understandings for cosmology and the nature of the universe.

Chapter 15. Entanglement. Einstein called this "spooky action at a distance" and doubted that it's a real phenomenon of the physical world. But experimental physicists have now confirmed that elementary particles can be so entangled that acting on one can immediately affect the characteristics of its far-distant entangled partner. The phenomenon of entanglement requires an explanation since it demonstrates communication of information faster than the speed of light—even instantaneously—in contradiction of a basic assumption of physics that caps transmission rates at the speed of light.

Chapter 16. Entropy and Information. Entropy is a thermodynamic property, which, perhaps surprisingly, is a measure of disorder. A basic principle of physics is that closed physical systems migrate in the direction of increasing disorder. This is one of the physical world's rare instances of an arrow of time. Modern physics links entropy with information content: high infomation content means low entropy. Theorists have linked these concepts with basic questions about the nature of time, the nature of thinking, and other phenomena of new physics.

Chapter 17. Black Holes. Science meets science fiction with this paradigm for the mysterious nature of spacetime. Black holes are linked in many directions to other new physics phenomena.

Chapter 18. Imaginary Time and Multiple Histories. Physicist Stephen Hawking has combined these two phenomena—the mathematical construct of imaginary numbers applied to time, and the parallel worlds of unobserved quantum phenomena—to propose a broad theory of particle physics and cosmology.

Chapter 19. Tunneling. The boundaries between adjacent matter are not as clear-cut as we might imagine: some particles can tunnel into what appears to be another particle's territory. This has practical applications and is also another element of new physics that raises questions about our theoretical understandings.

Chapter 20. Bose-Einstein Condensates. Quantum phenomena are generally thought to take place only at the smallest submicroscopic levels, but Bose-Einstein condensates are macroscopic quantum phenomena, which display a degree of coherence that had been considered possible only for elementary particles.

Chapter 21. Chaos and Complexity. Many contemporary mathematicians consider that complex phenomena cannot be understood as extensions of our normal mathematical frameworks, but must instead be described by a qualitatively different form of mathematics involving fractional and infinite dimensions as well as highly exaggerated consequences of apparently small variations in conditions. Physicists have applied the mathematics of chaos to various phenomena of the microscopic and galactic universes.

Chapter 22. Neutrinos. This elementary particle has been a thorn in the side of theoretical and experimental physicists. The standard model of particle physics assumes that neutrinos have zero mass, but experiment now confirms a small mass. This may have far-reaching consequences for physics theory.

Chapter 23. The Unreasonable Effectiveness of Mathematics. Scientists wonder why mathematics works so well as a tool of physics and other sciences. There seem to be implications for the nature of reality.

Chapter 24. The Myths of Time and Mass. Some radical understandings of the physical world propose that time is an illusion and a superfluous concept. And, surprisingly, physicists still feel that it is not clear why matter has mass: the standard model of particle physics proposes that mass derives from a never-observed elementary particle.

Chapter 25. The Role of Art. Some scientists propose that art is intimately linked with the deepest natures of the cosmos and the submicroscopic world. And concepts of modern physics find their way into modern literature and art. Physics is also linked to philosophy, religion, and paranormal phenomena.

Chapter 26. The Fine Structure Constant. Physicists wonder about the smallest units of space, time, mass, and force. There has been a numerologic obsession of sorts among some physicists involving the number 137.

12.
Quantum Gravity

Under general relativity theory, which is a classical not a quantum theory, the force of gravity is propagated by gravitational waves, which transmit the force of gravity at the speed of light.

Under a quantum theory of gravity, we focus on the quanta of gravity—gravitons, the elementary force particles that transmit gravity through a process of graviton exchange. Gravitons, never experimentally observed, are particles of zero mass which travel at the speed of light and have a quantum "spin" of 2.

Quantum gravity is of interest both to permit gravity to be unified with the other three forces, as well as to unify general relativity with quantum physics. Quantum gravity is at the heart of physics' Theory of Everything.

Without migrating our understanding of gravity into the framework of quantum physics, general relativity will remain a theory of classical physics, of physics not reconciled with quantum theory. This seems inelegant, that these two major theories of modern physics—general relativity and quantum physics—are not unified into a single theory. The quest for a convincing theory of quantum gravity is physics' search for the Holy Grail: quantum gravity will explain mysteries we're aware of and also answer questions we don't yet even know how to ask.

Some physicists, including Roger Penrose, who theorize about the connection between the brain and the mind, look for this connection at the intersection of general relativity and quantum physics. For these physicists, "everything"—in the Theory of Everything, for which quantum gravity is the centerpiece—must include not only force particles and matter particles, not only general relativity and quantum physics, but also a theory of consciousness.

Lee Smolin's 2001 book *Three Roads to Quantum Gravity* draws from current approaches to propose his preferred approach to quantum gravity. We'll start our discussion with Smolin's approach to this central question of physics that remained unanswered as the new millennium began.

***

Background Independence

The three roads that Pennsylvania State University physicist and geometer Lee Smolin sees physicists taking toward quantum gravity are string theories, loop quantum gravity, and black hole thermodynamics. Smolin draws from these three approaches to recommend how quantum gravity is to be best understood.

String theorists are working very hard to create a theory of everything out of models in which strings are the elementary entity of physics. Today's extradimensional string theories have proven remarkably robust in creating accurate models of physics' particles and interactions. String theories provide a model for all four of physics' forces and for the elementary particles of matter as well. Among string theories' vibrational string patterns is a pattern that exactly produces the properties of the graviton. Thus, string theory is a quantum theory that incorporates gravity.

String theories' detractors are concerned that new features seem to proliferate in order to respond to objections and to match the theory with observed reality. But it's hard not to be impressed with how all-encompassing string theories are. Admittedly, there's a "Christmas tree effect" as new features and twists embellish the theory to get it more accurate. But it's a Christmas tree, not a national forest—it's still a fairly compact theory considering the scope of the questions it's addressing.

But string theories are background-dependent. The background is spacetime, and in string theories all of the forces—including gravity—operate against the background of spacetime. This seems to present a conceptual stumbling block in the way of using string theories to reconcile general relativity's gravity with quantum physics' other forces, because general relativity is a background-independent theory. Under general relativity, the force of gravity shapes spacetime—is spacetime.

So it's hard to see how a string theory road to quantum gravity can be the whole road: even if string theory's models provide extraordinary accuracy to the proposed structures of physics' forces and matter, background-dependent theories will not give us the gravitational exceptionalism that we need. Gravity is not like the other forces. The other forces operate against the spacetime backdrop, in a spacetime grid. Gravity doesn't. Gravity is the spacetime grid.

Theories of the nongravitational forces can be developed as background-dependent theories—the electromagnetic, weak, and strong forces operating against a background of a fixed spacetime grid. Theories of gravity also can (and have) been developed as background-dependent theories, but it's hard to see how a background-dependent theory of gravity can be correct at the deepest levels. Gravity—the force being described—reshapes spacetime. How can a theory of gravity be validly constructed assuming fixed spacetime when we know there's this recursive effect of gravity changing spacetime?

Smolin discusses the kind of radical rethinking that will be required to establish a complete theory of quantum gravity. The "three roads" of Smolin's title will ultimately have to combine to produce a single more basic theory, and this more basic theory will have to be background-independent. In fact, there are string theorists today who are working to reformulate string theory to remove its inherent background dependence.

What this single theory will ultimately be based on will in all likelihood move us entirely away from a conceptualization that the universe consists of things occupying regions of space, toward instead a conceptualization that the universe consists of a network of relationships, of processes by which information is conveyed from one part of the universe to another.

Loop Quantum Gravity

String theories assume that space and time are granular, that there is a smallest granule of length (the Planck length, 10^{-33} centimeters) and a smallest granule of time (the Planck time, 10^{-43} seconds). Said another way, "quantum geometry is discrete."[66] Of course, these granules of space and time are so unimaginably small that there is no experimental verification that the geometry of spacetime is in fact discrete rather than continuous.

Loop quantum gravity, the second of Smolin's "three roads to quantum gravity," also assumes this granularity of space and time. But loop quantum gravity, of which Smolin is a key developer, sets out as its quest to remove background dependence. An earlier version, a lattice theory of gravity, was rejected, because the background dependence could not be eliminated. But loop quantum gravity does this by reducing spacetime to loops alone, with no background in which these loops reside. The loops interact in a network of "knots, links and kinks."[67] Taking the lead from work that Roger Penrose had done in the 1960s and 70s, the loop quantum gravity model is a spin network model, because the loops each have a value associated with the spin of physics' elementary particles. The surface and volume of spacetime are built up at the edges and nodes of this spin network. Because spacetime at the Planck scale is not localized at a point, these spin networks have come to be called spin foam, and are currently a subject of exploration and theorizing among physicists across the world.

What's Going On With . . . Spin Foam

Grosse and Schlesinger[68] propose a spin foam model of M Theory, which is today's umbrella theory unifying the various strands of string theory. As a spin foam model, this work proposes a background-independent version of string theory, thought to be critical for permitting string theory's full reconciliation of general relativity and quantum physics.

And a number of physicists[69] note the success of spin foam models at unifying multiple approaches to quantum gravity. These physicists all present refinements to the spin foam approach, most of which are variations of the Barrett-Crane model developed by mathematicians John Barrett of the University of Nottingham and Louis Crane of Kansas State University.

Smolin's Resolution, and Its Critics

Smolin favors loop quantum gravity over string theories as the source of a theory of quantum gravity, due to loop quantum gravity's inherent background independence. But Smolin also recognizes the power of string theory for describing forces and matter, even though this is accomplished against a classical spacetime background. For this reason Smolin views loop quantum gravity and string theory as complementary and ultimately reconcilable (with loops forming a more basic concept than strings). Perhaps, too, string theory will one day be successfully reformulated (as Grosse and Schlesinger propose) as a background-independent theory.

Smolin also discusses how both loop quantum gravity and string theories interrelate with black hole thermodynamics, the third of Smolin's three roads to quantum gravity.

Smolin estimates that there are a billion billion (10^{18}) black holes in the universe. Those looking to create a theory of quantum gravity from the study of black holes do so by incorporating concepts of the black hole's entropy (degree of disorder, a key metric of thermodynamics) and the amount of information that a black hole—or any region of space—can contain.

For example, University of Oxford physicist Stephen Hawking theorizes (*A Brief History of Time* and elsewhere) that a quantum theory of gravity is needed to understand how the universe began: without a quantum theory of gravity, theories produce at the big bang an undesired collapse of all matter to zero volume, an undesired infinite density and infinite curvature of spacetime at the beginning of time. By drawing conclusions from studies of black holes, Hawking has proposed a theory of the universe, which includes quantum gravity, based on the concepts of imaginary time and the universe's "multiple histories."

These concepts—entropy, information, black holes, multiple universes, imaginary time—will be discussed in upcoming chapters. In his stab at reconciling this black hole thermodynamics "road" with the other two roads to quantum gravity, Smolin proposes that that the holographic principle—which quantifies how much information can be contained in any region of space—will be a basic principle that unites the three roads to quantum gravity. (Physicist Dennis Gabor won the 1971 Nobel Prize for physics for earlier work on the holographic method.) Under this unified approach, the universe is "a network of holograms,"[70] a network of information.

Although we will be pursuing further elements of these concepts, it is probably not a surprise that the work of Smolin and colleagues studying quantum gravity is not universally accepted. For example, Michael Riordan—who teaches the history of physics at Stanford and at the University of California, Santa Cruz, and who has written the general-audience book *The Hunting of the Quark*—expresses a number of criticisms of this pursuit of quantum gravity. Principally, Riordan's concern is the lack of experimental verifiability of these theories, the risk that "these imaginative theories of quantum gravity will remain rooted only in the misty realms of metaphysics." In Riordan's view, this contradicts four centuries of scientific method, and exposes physics to the "virulent attacks of postmodernist critics, who argue that that science has no special claim to objective reality." Riordan asks, "Are these people really practicing science?"[71]

Why Is This So Hard?

Why is it so difficult to construct a complete theory of quantum gravity?

If your junior high school or high school science courses were anything like mine, somewhere along the line you had a lesson on Galileo dropping objects from the Leaning Tower of Pisa. You were fooled into saying that a heavy ball would fall more quickly than a lighter ball, at which point your science teacher told you that, no—they'd both hit the ground at the same time. This bothered you, because it seemed so counterintuitive. Then you were stunned into silence by the claim that—in a vacuum, without air resistance—even a feather would fall at the same pace as a lead ball.

Rather than this being something that fourteen-year-olds should feel foolish about not understanding, this is actually an important mystery of physics. We expect, when we have more of a force-generating quantity, to have greater force. A big light shines more brightly than a small light. A nucleus with lots of protons has more strong force holding it together than a nucleus with few protons. So why would a more massive block of matter not fall faster than a less massive block—doesn't more mass mean more gravitational force?

The answer, in its general form, has to do with the precisely offsetting phenomenon of inertia: the larger the mass, the more force it takes to accelerate it. This question—the question of why inertial mass is equivalent to gravitational mass—is not at all a trivial question of physics, and in fact remains one of the mysteries of gravity. Attempts to solve this mystery have led to some strange hypotheses, such as University of California physicist Shu-Yuan Chu's recent work suggesting that this involves forward- and backward-in-time interactions between objects on earth and all of the other matter in the universe.[72]

And this mystery about the nature of gravity is just one element of why it is so hard to establish a theory of quantum gravity. Also noted as sources of difficulty in successfully modeling quantum gravity are:

> Gravity is so weak that the effects of quantum gravity are noticeable only at the smallest scales of distance.
> Gravitons interact with everything—all forms of matter, all forms of energy (remember: matter and energy are equivalent, through $E = mc^2$), even with other gravitons.
> As we've mentioned, spacetime plays an active, dynamic role in gravity, unlike the passive role it plays as the stage on which the other forces act. [73]

In the meantime, experimental physicists aren't standing still while the theoreticians work this out. At both subatomic and astronomical lengths, physicists continue the search for evidence of the graviton and quantum gravity.

What's Going On With . . . Finding Gravitons

At the subatomic level, the possibility of extra spatial dimensions besides our familiar three could result in quantum gravitational effects occurring at low enough energy levels to be observable by our next generation of particle colliders. Physicists are preparing for this possibility by discussing methodologies for distinguishing collider evidence of gravitons from what would be effects of other "new physics" sources.[74]

And at the astronomical scale, physicists are exploring the possibility that ultrahigh-energy cosmic rays, interacting with our atmosphere, could provide our first "signature" of gravitons (as well as of large extra dimensions). One possibility is that horizontal air showers (which have never been observed, but which could result from cosmic rays crossing our atmosphere at an angle that carries them across a longer horizontal route) could result in successful observation of gravitons in our skies.[75]

In addition, many physicists are now looking to develop a satisfactory theory of quantum gravity only as a result of radical rethinking of current notions of geometry, logic, philosophy, or causality. Applying quantum physics to the cosmology of the big bang, some propose that both matter and geometry were simultaneously created "from the vacuum of a flat, empty space-time without structure."[76]

***

A 2001 progress report on contemporary approaches to quantum gravity notes: "Work along these lines has not yet led to any physical breakthroughs, but perhaps that is too much to ask, given that more conventional approaches have not been terribly successful either." But the analysis continues: "It is safe to say that most people working in quantum gravity expect that the theory will eventually lead to radical changes in our understanding of space and time."[77]

13.
Extra Dimensions

Extra dimensions, beyond our familiar three dimensions of space plus a time dimension, are currently a subject of serious—even intensive—investigation by physicists. And not just by theoretical physicists, who are modeling how various possible configurations of extra dimensions impact both the subatomic world and the large-scale universe. Experimental physicists, too, are researching extra dimensions, in order to actually observe extra dimensions of space.

***

Extra dimensions are an established presence in science fiction and other entertainment and art. The concept of extra dimensions intrigues us—strikes a resonant chord that rings true enough to hold our interest and form a plausible basis for speculation.

In his 1994 book *Hyperspace: A Scientific Odyssey Through Parallel Universes, Time Warps, and the 10th Dimension*, City University of New York physicist Michio Kaku discusses the longstanding interest in mystics' and philosophers' speculations on "hyperspace"—those extra dimensions "tantalizingly close, in fact surrounding us and permeating us everywhere we move, yet just beyond our physical grasp and eluding our senses."[78]

Kaku cites examples of extra dimensions in literature, music, art, psychology, religion, and politics. H. G. Wells' *The Time Machine* and Lewis Carroll's *Alice in Wonderland* are examples, and so are Pablo Picasso's impossibly configured faces painted with both eyes looking at us from just one side of the face. Extra dimensions have been hypothesized as the locations of heaven and of hell and of the Bermuda

Triangle. Star Trek and Star Wars take advantage of extra dimensions for interstellar travel and for plot enhancement.[79]

In fact, the mainstream world of physics had an important brush with the fourth spatial dimension during the 1920s, when two physicists—Theodor Kaluza from Germany and Oskar Klein from Sweden—proposed a unification of the two forces then known, gravity and electromagnetism, but in five dimensions (four spatial dimensions + time). Actually, Kaluza and Klein never worked together on this, with Kaluza publishing in 1919 his part of what's now called the Kaluza-Klein model, and Klein incorporating this into a more complete theory in 1926. This theory went a significant way toward unifying all of the physics then known—relativity, quantum physics, and electromagnetism and gravity, which were the only two forces then known—by showing that electromagnetic radiation can be modeled as a ripple in the fifth dimension, as Einstein had showed that gravity is a ripple in the fourth dimension.[80]

A brief comment on numbering dimensions: time being considered the fourth dimension achieved serious scientific status earlier than hypotheses about extra spatial dimensions beyond the familiar three. So, unless context makes clear otherwise, when physicists refer to four dimensions, they mean three spatial dimensions plus time (sometimes shortened to 3 + 1). And when they refer to five dimensions, they mean 4 + 1. But do keep your mind open to the context: four or five dimensions can be a reference to four or five spatial dimensions, with time being in addition, if the context makes this clear.

Clearly, any extradimensional hypothesis faces an immediate burden of explaining why we don't perceive the extra dimensions in our normal day-to-day existence. As does much of subsequent theorization of extra dimensions, the 1920s' Kaluza-Klein theory assumed that the fifth dimension is not seen because it is so compact that it is smaller than what our senses, even enhanced by all known technological aids, can pick up.

The 1920s' Kaluza-Klein theory had some important inconsistencies that were not resolved. But another reason for this theory's not having become a sensational theory of everything is that two additional forces of the universe—the strong and the weak forces that operate within the atom—were subsequently confirmed, and this gave mainstream physics four decades of a different direction, during which the physics of the atom became the major focus.

Physicists have made great theoretical and experimental strides in observing and modeling the universe's elementary particles and how they combine to form atoms and molecules. String theory—which at its early phase was a theory of the strong force and the elementary particles—revived the interest in extra dimensions, since with string theory a variation of the original Kaluza-Klein models became a plausible model for modern physics' more expansive universe of forces and particles.

Today extra dimensions are sought and hypothesized in their own right and not just in service of string theory. Extradimensional models are still called Kaluza-Klein models, even though their applications have extended far beyond Kaluza's and Klein's own developments.

Where physicists today are headed with extra dimensions is very much in flux. For one thing, the theoretical basis for how many dimensions are being hypothesized has remained uncertain. As a result, many papers and analyses are presented based on an indefinite N extra dimensions. This has not stood in the way of progress, because the mathematical models are generally robust enough to proceed without N being specified: the value of N can be dropped into these models if and when consensus on its value is achieved.

The early mainstream of extra dimensions assumed compact extra dimensions, with gravity being the only force operating in the extra dimensions. Often the assumption was that the extra dimensions were only as large as the smallest quantum of length, the Planck length, which is 10^{-33} centimeters. The idea is that the spatial structure at the point of the big bang was nondiscriminatingly multidimensional, but at an extraordinarily early moment after the big bang only our three familiar spatial dimensions continued expanding, while the extra spatial dimensions remained compact.

Some subsequent thinking relaxed the size of the extra dimensions a bit. As long as the extra dimensions remained comfortably subatomic, they could be credibly modeled as invisible to us. Theorists have demonstrated that it doesn't require extra dimensions quite as small as the Planck length to explain their not having been observed.

Even more recently, some or all of the extra dimensions have been hypothesized as "large," but still not noticeable to us. However, here "large" means a few millimeters. The significance of this millimeter-magnitude length is that this is the smallest distance at which we have been able to probe the gravitational force. Below this distance, the gravitational force is too weak for any current instruments to be able to pick it up. In other words, these dimensions would be trillions of trillions of times larger than the Planck length—even larger than an atom—but still explainable as not yet detected.[81]

Some recent theories even propose infinite-size extra dimensions. In other words, the hypothesis is that these extra dimensions continued expanding after the big bang, as did our familiar three spatial dimensions, and therefore these extra dimensions are not compactified relative to the familiar three dimensions. Often, these hypotheses of infinite-size extra dimensions assume a curvature of the extra dimensions that results in effects similar to compactification. Or, alternatively, these hypotheses of infinite-size extra dimensions may assume that forces operating in the extra dimensions are strong only at certain dimensional intersections.[82]

The explanation for no detection of these larger or infinite extra dimensions generally requires a special role for gravity. There is significant momentum today behind a model in which only the gravitational force exists in the extra dimensions. We—along with the three forces of electromagnetism and the strong and weak forces, and along with just the weakest shadow of an omnidimensional gravitational force—exist on a three-dimensional wall of a larger extradimensional bulk. By contrast, gravity—and the geometry of space—are universal, extending to all dimensions.

This concept—that only a weak shadow of the gravitational force operates in our $(3 + 1)$-dimensional world—also has the advantage of explaining why, at least in our $(3 + 1)$-dimensional world, gravity is so extraordinarily weak. Gravity is ten trillion trillion (10^{25}) times weaker than the next weaker force, which is the weak force—weak only in comparison to the other two forces but much stronger than gravity. The weak force, in turn, is one hundred billion (10^{11}) times weaker than the electromagnetic force, which in turn is about one hundred times weaker than the strong force.

The gravitational force in particular just seems too weak to be included within a reasonable common theory with the other forces. Physicists refer to the *hierarchy problem*, noting the *desert* between the strength of gravity and the strength of the other forces. The extradimensional explanation is that gravity appears weak to us because its true larger strength has been masked through the dilution of its strength into the extra dimensions of the bulk.

There had been some serious investigation and theorizing (now largely dismissed) that the discrepancy in strength between gravity and the weak force is so large that there must be a fifth force, with strength in between. It shows how far extradimensionality has progressed into the mainstream that extra dimensions are now considered more credible than a fifth force.

Particle Accelerators and "TeV Physics"

Particle accelerators, an important investigatory tool for particle physics, allow experimental physicists to observe particles' subcomponents by smashing very fast-moving particles into each other and into walls.

The slowing down or stopping of these particles creates an additional aspect of this tool: it allows the creation of additional particles out of the "vacuum of space." These additional particles are not subcomponents of the smashed particles; they are different particles. They arise because the significant decrease in the smashed particles' velocity releases a large quantity of energy (velocity goes down, so energy goes down), which (quantum physics tells us) becomes available to convert to mass, according to the relationship $E = mc^2$. Since c (the speed of light) is so large, and c^2 is even larger, only small amounts of mass m are produced by large releases of energy E. So these particles may be very small and short-lived.

Additional amounts of mass can be produced if the smashed particles are collided with their antiparticles. Then mass loss resulting from particle/antiparticle annihilation means extra amounts of energy become available.[83]

Moving backwards in time from today back toward the big bang—when the universe was dramatically smaller, hotter, and more energetic—1 TeV (a trillion electron volts) was the energy level at which the electromagnetic and weak forces were unified. Before this time, when the universe's energy level was even higher, the electroweak force existed as a single unified force. After this time, as energy levels dropped, the electromagnetic force and the weak force became distinct.

TeV means tera, or 10^{12}, electron volts. This is a measure of energy, and is the energy level of electroweak unification. Traditionally (before consideration of extra dimensions), grand unification—the unification of the electroweak and strong forces—was at an even earlier and more energetic time, at 10^{25} electron volts. And, in this nonextradimensional framework, we find our theory of everything—all forces, including gravity, united—at 10^{28} electron volts.

The next generation of particle accelerators, being built today, will be capable of exceeding the TeV level of energy. So excitement has been building for the full level of experimental verification that we will see for the electroweak unification.

But the possibility of extra dimensions has changed all this. Extra dimensions are theorized to permit the unification of all four forces at the TeV range of energy: since we've been seeing only a weak shadow of gravity, on our three-dimensional wall, it's appeared to need a hierarchically discontinuous boost in energy to become unified with the other forces. But with gravity assumed to exist at greater strength throughout an extradimensional bulk, full unification may take place in the range of the electroweak unification energy level, which is around the corner as the capability of our experimental observation.

The energy level—10^{28} electron volts, in the traditional, nonextradimensional framework—at which gravity is unified with the other forces is energy of Planck scale. And the discrepancy between the Planck scale and the scale—10^{12} electron volts, or 1 TeV—of electroweak unification reflects the hierarchy problem, the desert of 16 orders of magnitude between energy of Planck scale and energy of electroweak unification. Theories of extra dimensions propose to narrow or eliminate this desert—that is, to solve the hierarchy problem—by eliminating the problem. This exciting theoretical possibility offers the possibility of experimental verification of the unification of all forces, including gravity, with our next generation of particle accelerators.

What's Going On With . . . Finding Extra Dimensions

Experimental physicists use several different methods to look for evidence of extra dimensions.

One method is to try to find evidence of gravitons escaping into the bulk. To do this, physicists look for otherwise unaccounted for losses of energy, which match the amount of energy loss that would result from gravitons being carried off into dimensions that are not accessible to our measurement.

Other physicists are looking not to gravitons, but to the other force-transmitting particles. For example, in the standard model the photon, which transmits the electromagnetic force, is massless. But some hypothesize that the photon propagates into the bulk with mass but without carrying the electromagnetic force, which remains confined to the wall. They then look for evidence of energy loss that corresponds to the inaccessible photon mass that is carried into the bulk.

Still others look for previously unconsidered patterns of new particle creation, which had not been considered under the standard model. For example, some physicists contemplate the implications of new forms of extradimensional graviton-graviton exchange.[84]

1 TeV = 1000 GeV

There has been some thought that we won't have to wait until 1 TeV (1 trillion electron volts) particle accelerators come on line before we see direct evidence of extra dimensions. This is because another direction of theorizing suggests that perhaps the strong, weak, and electromagnetic forces are not fully confined to the three-dimensional wall. If in fact some or all of the extra dimensions are "universal extra dimensions" (that is, not just gravity's extra dimensions), then we may see the as-yet-unseen grand unification of these three forces when we reach energies of a few hundred giga electron volts (1 GeV = a billion electron volts). Since today's accelerators can reach energy levels of about 100 GeV, this grand unification from universal extra dimensions would not yet have been observed.[85]

Traditional grand unification models make no reference to gravity,[86] in spite of the traditional grand unification energy (10^{25} electron volts) being not much less than the traditional Planck scale (10^{28} electron volts) at which gravity is unified with the other forces. In other words, the desert between the grand unification and electroweak unification energies is almost as large as the desert between the Planck scale and the electroweak unification energy. Bringing the grand unification energy down to the level of electroweak unification is actually a problem that is independent of the problem of lowering the Planck scale to the electroweak unification level,[87] but one with equally exciting prospects for confirmation and investigation by particle accelerator experimentation.

There is significant excitement among scientists as to the new phemonena that will be observed and the conclusions that will be drawn as experimental physicists continue these particle accelerator studies.

What's Going On With . . . Theories of Extra Dimensions

While we wait for more powerful particle accelerators to come on line and enhance our experimental capabilities, there's time for additional abstract theorizing.

Physicists have shown that variations in the shape of extradimensional space can result in effects that are similar to the effects of varying the size of the extra dimensions or the number of extra dimensions.[88]

Some physicists speculate on crystal universes of many dimensions,[89] and others speculate on the implications of one hundred or even an infinite number of extra dimensions.[90]

Castro investigates the possibility of an infinite number of dimensions to spacetime and frames his analysis within the fractals and chaos theory fields of mathematics, fields that have attracted both enthusiastic proponents and dismissive detractors. Castro sees broad implications of infinite-dimensional spacetime, transforming to observer-dependent significant aspects of what more established physics has considered fixed or constant. In Castro's fractal spacetime, the speed of light, Planck's constant, the gravitational constant, and the number and structure of spacetime dimensions all depend on the observer and are not fixed.[91]

Later in this book, as we look at how various physicists have put together many of the speculative topics we've discussed, we'll be returning to speculative frameworks in which the structure of spacetime is actively created by the observer, by the conscious mind, and in which information plays a central role in structuring the physical world.

A Universe of Many Folds

Arkani-Hamed, Dimopoulos, and Dvali are three early developers, back in 1998, of the possibilities for gravity extending its reach into extradimensional space. Their hypotheses are often abbreviated as the ADD hypotheses.

These three physicists, along with Kaloper, published a 1999 article on the "Manyfold Universe." This article proposed that our (3 + 1)-world is "folded" many times over inside extradimensional space. Extradimensional geometry such as this would have numerous interesting implications for many of physics' open questions, including questions we'll be addressing in upcoming chapters. For example, objects that appear very distant as observed and measured in our (3 + 1)-world may be gravitationally very close, if the "folds" brought them nearby in the bulk. This could lead to an explanation for "dark matter" (discussed in the next chapter), which has mass but can't be seen.

We'll also be discussing superluminal phenomena—phenomena which appear to be transmitted faster than the speed of light, in apparent violation of a basic principle of relativity. These phenomena too could be explained by gravitational transmission that is a short distance in the bulk but a long distance in the folded-over (3 + 1)-wall which forms the spacetime in which we (and the standard model) operate.

***

We will see extra dimensions again and again, as proposed explanatory frameworks for many of Part Three's phenomena of new physics, as well as for some of Part Four's theories of new physics and the mind.

14.
The Universe

There is a long history of humanity's trying to figure out our place within the universe. It's fairly recent in human history that we've moved away from thinking of ourselves as the universe's center. And as we've found out more about the scope of the universe, our place in it moves farther and farther from appearing to have any special centrality or significance.

How big is the universe? Will it grow forever? Shrink? Remain static?

These great questions of cosmology still have no generally accepted answers.

It is somewhat surprising, actually, that we can't figure out which direction the universe is headed. The difficulty arises from how close a call this is. This difficulty is called the "flatness" problem, which results from the apparent growth rate of the universe being so very close to the dividing line between a universe that is curving outward and will grow forever vs. one that is curving inward and will shrink.

A "flat" universe is the shape of the universe at this dividing line, and it is the average density of the universe's matter that will determine the universe's ultimate fate. If the universe's density is at the critical dividing line, called "omega = 1," we're heading toward a flat universe, slowly decelerating toward a zero-growth structure. If the universe's density is higher than this critical level, gravity will eventually overtake the outward force of the big bang, and we'll be headed toward the "big crunch" of a shrinking universe. But if the universe's density is less than this critical level, the "big freeze" of eternal expansion will be the universe's future.

Today's observations of the universe's average density of matter do not allow us to determine with certainty whether we're headed for a "big freeze" outcome, a "big crunch" outcome, or a flat universe. This is spite of the very narrow range in how today's observations roll back toward conclusions about how the universe unfolded after the big bang. The range of uncertainty in today's measurements of the universe's density equates to a variation of one part in a quadrillion (10^{15}) in the universe's density one second after the big bang. As you might conjecture, this one part in a quadrillion uncertainty is hovering right around omega = 1. Based on today's observations, we know that one second after the big bang, omega fell somewhere between 0.999999999999999 and 1.000000000000001.

How can the universe be so fine-tuned that this is such a close call?

***

Einstein's Greatest Mistake

Albert Einstein called "my greatest mistake" his invention of a "cosmological constant" to answer an early form of the question of the universe's shape.

There was no theory of the big bang when Einstein postulated the cosmological constant, and Einstein believed that the universe was, is, and will be a large and static space. No astronomical observations had yet taken place to confirm what we know today about the expansion of the universe. In fact, the problem that Einstein faced was that it was hard to understand why the universe wasn't shrinking: gravitation seemed to be the dominant force of the far reaches of the universe, and gravitation should cause the matter within the universe to pull in towards the universe's center.

In order to construct a theory of a static universe—one that is not shrinking—Einstein postulated (with no evidence to support it, but with only his conviction that the universe must be in a steady state) that a cosmological constant must exist as a measure of a basic force of nature that pushes the universe outward, against its otherwise shrinking gravitational tendencies.

By the late 1920s, American astronomer Edwin Hubble and his colleagues had convinced the scientific world that the universe was in fact expanding. They did this by observation of the "red shift" of the light of distant galaxies, which indicates that the galaxies are moving away from us. At first, some thought that this was just further confirmation of the Earth's being the center of the universe—why else would everything, in every direction, be moving away from us? But others realized that actually everything in the universe is moving away from everything, similar to all points on the surface of a balloon getting farther away from each other as the balloon is blown up.

Hubble's analysis gained widespread acceptance, although convincing experimental verification was not published until 1965. Two Bell Lab scientists, Arno Penzias and Robert Wilson, were disturbed by observation of some background radiation that just did not seem to go away, even after all earthbound, solar-system-based, and Milky Way sources were eliminated as possible causes. Penzias and Wilson were winners of the 1978 Nobel Prize for physics for their discovery of this cosmic background radiation—radiation which emanates with equal strength everywhere ("homogeneity") and from every direction ("isotropy"), as a remnant of the big bang.

So Einstein's "greatest mistake" was really two-pronged: the universe is expanding, not static, and we don't need a cosmological constant to represent a hypothetical new force to cause expansion—expansion results from the big bang.

With expansion established, this led to a natural question: will the universe expand forever? Or will there be some point at which the inward gravitational force will overtake the outward pull that originated with the big bang, resulting in a reversal towards a "big crunch"? It is because of the flatness problem that these questions are so hard to answer.

The Horizon Problem

Cosmologists also wondered about the uniformity—the homogeneity and isotropy—of phenomena such as the big bang's remnant background radiation. This radiation results in an almost precisely uniform background temperature in the vast reaches of space of 2.7° centigrade above absolute zero (called 2.7° Kelvin).

As the stages of the big bang are mapped out, we know that this radiation results from photons' interactions with atoms, which couldn't have begun until there were atoms, which first developed 300,000 years after the big bang.

Today, we are receiving cosmic background radiation deriving from this event 300,000 years after the big bang. We can look toward the east for this radiation and find its resulting temperature to be 2.7° K (2.7° Kelvin). We look toward the west and also find the temperature of the cosmic background radiation to be 2.7° K. Why is this cosmic background radiation so uniform? How was there a common communication of the physical conditions for the radiation traveling to us from the east and the radiation traveling from the west? This has been likened to "finding two ancient civilizations on opposite sides of the Earth with nearly identical languages. The civilizations must have been in causal contact."[92]

Under the standard big bang theory (before the theory of "inflation" was developed), it can be calculated that a communication creating these common physical conditions would have to have traveled at one hundred times the speed of light in order for this uniformity of physical conditions to have been established—that's the speed required to communicate universe-wide, based on the size of the universe 300,000 years after the big bang.

Since any type of communication faster than the speed of light violates a basic principle of special relativity, the standard big bang theory just assumes that uniformity of conditions continued after the big bang, without providing any mechanism for the processes and parameters of this uniformity to have been communicated.

Under the standard big bang model, the size of the universe—at 300,000 years after the big bang, at 200,000 years, all the way back to its smaller and smaller sizes at earlier and earlier moments closer to the big bang—was always too big for a uniformity of conditions to have been created subluminally (at less than the speed of light). At any point in time, the *horizon distance* that could be traversed at the speed of light was less than the radius of the universe.

Said another way, matter separated as much as the diameter of the 300,000-year-old universe is *causally disconnected*. And (without inflation) the aggregate substance of the universe was causally disconnected all the way back to the big bang.

Rather than assume superluminal communication of information all the way out to the full radius of the universe, the standard big bang model simply assumes (without explanation) a uniformity of physical conditions out to the full radius of the universe. This is what creates the horizon problem, the problem of trying to offer an explanation (without superluminal communication) of how the universe exhibits a uniformity beyond the horizon distance.

Inflation creates this explanation by offering an alternative picture of the first few trillionths of a trillionth of a trillionth of a second after the big bang.

Trillionths of a Trillionth of a Trillionth of a Second

A huge amount of theorizing has gone on regarding a march back in time to the big bang. Long before the universe's stars existed, there was a hot cosmic foam. And before that, the universe was so hot that the forces were all one, forces and matter were not distinct, and perhaps there were more than three equally realized spatial dimensions.

At one tenth of a trillionth of a trillionth of a trillionth of a second after the big bang, the (then very small) universe temporarily expanded very rapidly—faster than the speed of light. This is the concept of inflation: for about ten trillionths of a trillionth of a trillionth of a second—from the period 10^{-37} of a second after the big bang to a period 10^{-35} of a second after the big bang—the universe expanded 10^{23} times faster than its previous or subsequent pace. This is also referred to as a supercooling, since for cosmological models expansion goes hand in hand with energy decreases and temperature drops.

So unlike the standard (no inflation) model of the big bang, the inflationary model creates the opportunity for subluminal communication of the rules of the universe's uniformity. This uniformity is established universe-wide in the the period after the big bang up until 10^{-37} seconds after the big bang. The inflationary big bang model (but not the standard big bang model) allows this because the universe is much smaller during the first 10^{-37} seconds after the big bang under the inflationary model than it is under the standard model. Thus inflation solves the horizon problem, including its homogeneity and isotropy aspects.[93]

What Other Problems Does Inflation Solve?

Inflation was first hypothesized in 1981 by Alan Guth, a young Stanford physicist, later at M. I. T. It has gained wide acceptance for how it has solved a number of problems of modern physics. Hundreds of articles are published each year in journals of physics, astronomy, and cosmology about the implications of inflation.

Inflation solves the flatness problem by its very burst of expansion. Similar to a balloon losing its wrinkles as it is quickly blown up, the surface of the universe is flattened out after inflation. Of course, the mathematical proof of this is more elaborate, but Guth uses this balloon metaphor to help explain the creation of a flat geometry.[94]

Inflation also solves a more theoretical problem of physics, related to *primordial monopoles*. Ordinary magnets have a north pole and a south pole, and if you break a magnet in two you get two magnets, each also having a north pole and a south pole. But the Grand Unification Theory uniting electromagnetism and the weak and strong forces predicts that magnetic monopoles (with only a north pole or a south pole, but not both) should exist and should have been created at the big bang (that is, primordially). Monopoles have not yet been observed (our particle accelerators are not yet powerful enough), but before inflation there had been skepticism about why their predicted significant numbers and mass would not have slowed down the predicted rate of the expansion of the universe resulting from the big bang. This is where inflation comes in: it provides the acceleration to counteract the predicted retarding force of the primordial monopoles.

Another problem that inflation solves is the universe's clumping (this is also called the *inhomogeneity puzzle*). [95] Although at large scales the universe is extraordinarily smooth, there are certainly pockets of matter—including galaxies and supergalactic clusters—enormous relative to the size of Earth or the solar system, but nevertheless small on the cosmic scale of the universe. Under the inflation model, these derive from quantum effects (density perturbations) as inflation ends; these perturbations have grown to cosmic scales as the universe has expanded for the past fourteen billion years. Alternatively, the timing of these perturbations has been linked to quantum field fluctuations during the pre-inflation period[96] or during inflation,[97] in either case implying that both inflation and post-inflation expansion stretched the perturbations' effects to cosmic scales.

And how inflation ends is another problem that must be solved, albeit a problem of inflation's own making. (Remember that inflation ends just ten trillionths of a trillionth of a trillionth of a second after it starts.) This problem—called the *graceful exit problem*—has been solved by modeling of quantum tunneling, a kind of bubbling. Today's most commonly accepted explanation assumes a single "bubble" which the entire universe lies deep inside.

Most grandly, inflation is a theory of the big bang: it offers an explanation of how the big bang took place. Said another way, this is an explanation of how the universe began. Again, the standard big bang model does not offer an explanation for this; it accepts the big bang as a given (an assumption). But the explanation offered by inflation constructs a model in which a "false vacuum" expands rapidly, distorting space and creating a wormhole to a bubble wall in which a universe is born. In this model, there is a larger original universe in which it appears that a black hole has been created; that black hole represents the big bang creation of a new "child universe" such as ours.

Physicists today are looking to create consistency among a number of observations and hypotheses—the age of the oldest stars and of the universe, the rate of expansion of the universe (called the Hubble constant), whether in fact there is a cosmological constant. Some call this the "oldness problem"; Guth calls it the "age crisis" and says: "The age crisis is probably the most pressing unsolved problem in cosmology at this time."[98] While inflation theory does not offer a broadly accepted solution to the age crisis, it does offer plausible components of an explanation, which may prove true.

Perhaps the most startling implication of inflation derives from the fact that the inflationary model assumes that what became our universe was much smaller—during the period after the big bang but before inflation began—than what the standard (noninflationary) model assumes for this period. This is in spite of the fact that after inflation ends, both the inflationary and the noninflationary models assume our universe to be the same size, expanding at the same pace.

What this means is that, under the inflationary model, the observed universe is only a miniscule part of the entire universe.

The reason for this conclusion is that the inflationary theory assumes that today's observed universe is the same size as the standard theory's universe, but only the inflationary theory's universe was subjected to an earlier period of enormous rapid inflation. This rapid inflation, greatly faster than the speed of light, created an observational barrier, assuming observation is limited to the speed of light, between our observed universe and a vast nonobservable universe beyond.

When we trace backward in time today's observed universe of the inflationary theory, it must have been only a small part of a larger universe. This larger universe was fully subjected to inflation, and therefore its heritage today is an enormous unobserved universe that envelops the tiny observed universe in which we live.

"In the future we shall know more," notes physicist and geometer Lee Smolin.[99] Just standing still on Earth, every day we know more. This is because each day (each second, each nanosecond) we are bathed in new matter, energy, information from beyond what could previously have reached us traveling at the speed of light from elsewhere in the universe. We don't know what we'll find out tomorrow—today it is out of reach, because information comes to us no faster than the speed of light.

Physicists have speculated as to how long it will go on, that we will continue to know more. These speculations lead to implications about how long an intelligent civilization can expect to grow and survive.

Some warn that their best estimate is that we may have just two trillion years. Based on today's assumed rate of universe expansion, in two trillion years the last information from the still-expanding universe, traveling at the speed of light, will just be able to reach us as we continue to race off with the expansion. After that, information traveling toward us, even at the speed of light, will never reach us, and we will no longer be able to obtain information from new regions of the universe. And based on assumptions of higher rates of expansion, the permanent loss of our capability to obtain information about additional regions of the universe may even soon become detectable.[100]

Others have performed calculations based on ranges of rates

of expansion and concluded that parts of the universe are already permanently beyond our event horizon, out of causal contact with us. They calculate that eventually there will be a sphere with radius of 16.6 billion light years that will envelop us as our permanent event horizon. This will not occur for several tens of billions of years—more time than the universe has existed so far. But at that point, we will have permanently reached the end of new areas of the observable universe—we will have reached the end of what we can have effects on, and of what can have effects on us.[101]

But under *multiverse* models—such as those in which our universe is a child universe of a larger ensemble of universes—perhaps these restricted horizons are limiting the energy and information only for our local universe. Gravitational bags have been modeled, which look compact from outside, but inside contain matter and can proceed via quantum tunneling through a wormhole region to explode and create a huge universe.[102]

This leads to models of possibilities for supercivilizations much older than ours visiting or inhabiting our own causally connected domain, our inflated bubble of a local universe. These supercivilizations, considering "interbubble migration," would have to consider the matter and energy costs in constructing wormholes. Perhaps the optimal strategy would be to be to expend a lot of resources to complete a few wormholes quickly in order to colonize additional domains and then use their resources to build yet more wormholes and migrate further.[103]

Others model possibilities for the shape—the topology—of the universe, and propose that even a small universe can consist of multiply connected universes which our universe's topology has thus far hidden from us. They propose studies of cosmic background radiation to explore the possibilities of these other universes.[104]

Some of the vocabulary of today's cosmology looks five hundred years old, as variations on the "Copernican principle" are weighed. The Polish astronomer Nicholas Copernicus put forth that the sun, not the earth, is the center of the solar system. And today astrophysicists put forth models of "strong" and "weak" versions of the Copernican principle, in which the local topology of spacetime (that is, our observed universe) plays the role of Copernicus's earth, and the global topology (the entire universe, including that beyond our observation) plays the role of Copernicus's solar system. Under the twenty-first-century version of the strong Copernican principle, because our locally observed universe is not deemed to be special, we may conclude that the global universe looks just like what we can see. By contrast, under the weak Copernican principle we don't draw conclusions about the global from our local observations; this leads to a model of the global universe in the form of multiply connected topologically distinct local universes.[105]

The Role of Extra Dimensions

Some physicists explain inflation as a natural consequence of the splitting off of the strong force from the electroweak force at a very early moment after the big bang.[106] But it is natural for physicists to also consider how extradimensional models could account for inflation, and publication of ideas on this possibility began soon after Guth's 1981 publication of his theory of inflation. For example, an early proposal accounted for inflation by modeling a universe with approximately forty extra dimensions.[107]

More extradimensional ideas were proposed over subsequent years, along with an explosion of articles on other aspects of inflation. In a typical proposal, the contraction of extra dimensions, taking place very shortly after the big bang, created inflation in the three extended spatial dimensions.[108]

Physicists consider proposals more elegant when they do not, as some earlier proposals had, require "squeezing" of matter or stringy fluids or tensor fields from the extra dimensions to fuel inflation in the three extended spatial dimensions. In other words, the more elegant models are a "vacuum solution," based only on spacetime itself: the universe "creates its contents out of its own expansion."[109]

Other physicists look at extradimensional models of the universe's expansion to explain not only inflation but also the relative strength of physics' forces, possibilities for multiple periods of inflation, and even the big bang itself.[110]

Dark Matter

Our capabilities for astronomical observation have progressed so greatly that we are able to observe stars within galaxies beyond the Milky Way and precisely determine the speed with which these stars are orbiting about their galactic centers. What we conclude is that they're orbiting too fast for the amount of mass that we can see in their galaxy: at the speed of these stars' orbits, they should be flying out away from their galaxy, if the only gravitational force to hold them in is the force of the matter that we observe.

As an additional astronomical observation, we are able to observe whole galaxies within their supergalactic clusters. From these observations, too, we observe an apparent shortage of mass: the observed quantity of mass is too small to prevent galaxies from running away from their clusters.

Since we know that the stars within these spiral galaxies are in a continual orbit without renegade stars flying off, and since we know that galaxies are not stampeding out of their supergalactic clusters, we conclude that there must be hidden mass—dark matter—not currently observable to us, but nevertheless holding stars and galaxies within their orbits. The consensus today is that most of the universe's matter is dark matter.

No known particle fits the profile of dark matter. "The identity of dark matter is currently among the most profound mysteries in particle physics, astrophysics, and cosmology."[111] It's surprising how intertwined are the world of the small and the world of the large—how we can be looking to the microscopic world of particle physics to solve some of the mysteries of the giant worlds of cosmology and astronomy, and vice versa.

One of the challenges in creating a model of dark matter is that the matter must be stable enough to contribute reliably to the universe's quantity of mass. The neutrino is a stable particle, confirmed to exist. Whether it has zero mass or some small mass has been a stubborn mystery under long-term investigation. We'll be discussing later the near-consensus confirmation that has finally been achieved, concluding that neutrinos do in fact have a small mass—not large enough to have ever been detected previously. If so, because of the large number of neutrinos that were created in the early universe, they would account for at least some of the matter that needs to be accounted for—not dark matter, but a reduction in how much missing matter we still need to explain.

One general class of candidates for dark matter is WIMPs (weakly interacting massive particles)—matter that, although massive, has not been noticed by us because it does not interact to any noticeable degree with our means of observing or measuring mass. For example, under the supersymmetry particle model, each matter particle has a partner in the form of a (not yet discovered) force particle. The stability of the matter particle (for example, neutrinos) implies the stability of its supersymmetric partner (the neutralino), and these "partners" have been a focus of investigation as candidates for dark matter.

The search for dark matter is widespread. Particles are hypothesized, then the search is on to confirm they exist and have mass, and to explain why they have never been noticed. Q-balls, axions, and other hypothetical and confirmed particles from theories of supersymmetry, from the QCD (quantum chromodynamics) theory of the strong force, and from other theories are all under theoretical and experimental study, both in our familiar dimensions and in extra dimensions.[112] No convincing explanation has yet been developed for dark matter.

Dark Energy and the New Cosmology

The latest astronomical data suggest that the universe is expanding and that the expansion is accelerating. To account for this accelerating expansion, the "new cosmology" includes a significant effect from "dark energy," which has negative pressure—that is, it pushes the universe's matter outward, against the force of gravity. This dark energy—sometimes called quintessence—has the same effect as the cosmological constant that was Einstein's "greatest mistake," although the new cosmology assumes the presence of dark energy not for Einstein's purpose of creating a static universe, but rather for the purpose of explaining an accelerating expansion of the universe, which cannot otherwise be explained by known matter and energy.

A number of possible explanations for dark energy have been proposed, for example networks of topological defects, rolling and spinning fields ("quintessence and spinessence"),[113] and the influence of the "bulk" of extradimensional space. Some propose an explanation for inflation and the present-day cosmological constant as deriving from vacuum fluctuations in extra dimensions.[114] Others propose yet a different explanation relating to extra dimensions: that the acceleration of the universe's expansion derives from gravitation leaking into extra dimensions.[115]

An intriguing possibility is that Einstein's theory of general relativity permits gravity to be repulsive. This repulsive gravity has not yet proven to provide the explanation for dark energy, but "if the the explanation for the accelerating Universe fits within general relativity, it will be a major new triumph for Einstein's theory."[116] That would be quite a triumph for a theory almost a century old.

Ruth Durrer of the Department of Theoretical Physics at the University of Geneva, in her closing comments—"Frontiers of the Universe: What do we know, what do we understand?"—at a 2002 conference on cosmology, summarized the state of scientific investigation of dark energy this way: "I believe that the accelerating universe and the cosmological constant represents the biggest puzzle in present physics. There are diverse attempts to address it, but none of them could convince a majority."

With dark energy as an additional feature of the "new cosmology," let's take a census of the universe's aggregate mass and energy (remember that mass and energy are made equivalent by the relationship $E = mc^2$):

The visible universe's bright stars, planets, and hot gases account for only 0.4% of the universe's mass/energy content. Additional everyday matter—made of protons, neutrons, and other particles subject to the strong force, but not visible as stars—accounts for roughly 3.7% of the universe's mass/energy. This includes nonluminous matter caught in the intergalactic gases, and perhaps also in gases within galaxies or remnants of early generations of stars. Two components of this 3.7% may be of special interest: neutrinos constitute 0.1% of the universe's mass/energy content, and black holes constitute 0.04%.

The above categories mean that only 4% of the universe is matter as we have typically understood the term.

So the rest (96% of the universe) is strictly "new cosmology": roughly 23% dark matter and roughly 73% dark energy.[117]

Many scientists are energized by the opportunity to still get in on the ground floor: 96% of the universe is "dark" and poorly understood. But Durrer again captures the feeling of some: "It is frustrating that we have not made any progress in identifying the dark matter, in contrary, we have supplemented it by what we call the 'dark energy', a gravitationally and esthetically repulsive component which is even more mysterious than the dark matter . We have no understanding about the origin of these two most abundant energy components of the present Universe."[118]

Tachyons

Inflation takes place at superluminal speeds. Inevitably, this leads to thoughts of tachyons—hypothetical particles that move faster than the speed of light.

Tachyons have never been observed, but they are actually not ruled out by the theory of relativity. What relativity requires is that nothing traveling below the speed of light can ever increase its speed so much that it can cross the line and travel above the speed of light. But relativity would permit a tachyon, since it always travels faster than the speed of light. So tachyons could exist, although they'd be beyond our reach because there is no mechanism to cross the barrier of the speed of light.

In any case, something needs to be done to explain why the theory of inflation hasn't just substituted, for the horizon problem's troublesome need for superluminal information-sharing, an equally troublesome superluminal expansion of spacetime.

Why is it OK for spacetime to expand at inflation's speed, enormously greater than the speed of light?

Tachyons, which would operate only on the faster side of the speed of light, could be the key.

Within the past few years, physicists working throughout the world have proposed and refined models of inflation resulting from tachyon activity. These models incorporate tachyon matter, tachyon fluids, tachyon condensation, and rolling tachyons, all operating in our (3 + 1)-dimensional world as well as in extradimensional space.[119] Some propose the neutrino as a likely tachyon,[120] and others discuss the possibility of tachyonic matter playing a second role—not just a role in inflation, but also in the later cosmology of the universe as some form of dark matter.[121] The developing theories of tachyons' roles in inflation are not without their critics,[122] but physicists keep refining the theory, addressing problems with enhancements and with embellishments to the theory of how tachyons created inflation.[123]

Alternatives to Inflation

Extra dimensions have been looked at not just as an explanation for inflation, but also as an alternative to inflation—that is, an extradimensional model not assuming superluminal expansion. For example, some look at topology alone, and at extradimensional structures in and of themselves, not as causes of inflation, but to provide direct solutions for the horizon and flatness problems without a need to invoke inflation.[124]

Others have proposed an extradimensional alternative to inflation which they call the "ekpyrotic universe." This term is drawn from an ancient Greek cosmological model in which a fire—called ekpyrosis—cyclically consumes and recreates the universe. The ekpyrotic universe begins as infinite although empty, then creates the big bang by collision of our (3 + 1)-dimensional surface (our "3-brane") with another 3-brane that is moving within the extradimensional bulk.[125]

There have also been other proposals to solve these problems directly, without including inflation itself within the solution. Some, for example, hypothesize on the implications of time-varying speeds of light (faster in the early universe) as an alternative to inflation's superluminal expansion of the universe's spacetime.[126] And others propose the alternative to inflation of a variable fundamental mass (variable Planck mass), in which an early "massively aged and detained" epoch meant that the universe is actually older than the fourteen billion years of the standard model.[127]

Even though inflation may have become "cosmological dogma,"

some feel that the concept of the oscillating universe remains a viable alternative in solving the questions of flatness, horizon, and density perturbations. The oscillating universe proposal is an enhancement of physicist G. Lemaître's 1933 Phoenix Theory, and involves a series of big bangs and big crunches, continuing in ever longer cycles. The current cycle is the first in which the conditions permit the formation of galaxies, stars, and biological life. This model does not assume the standard inflation theory's period of superluminal expansion, and is offered as an alternative that does not require the substantial fine tuning of various factors, including the cosmological constant, which must be assumed in order to make the inflationary theory match observed facts.[128]

Roger Penrose, whose classification of history's theories of physics was discussed within Chapter 2's "report card" on twentieth-century physics, is not a fully convinced fan of inflationary theory. For one thing, Penrose makes the point that inflationary theory is based on assumptions of grand unified theories that Penrose views as unsubstantiated and tentative. But in addition, Penrose makes the point that inflation requires both a discontinuous beginning ("initial singularity") and a discontinuous end ("final singularity"), but introduces no "time-asymmetric ingredient" to explain what triggers, asymmetrically, both a transition from noninflationary to inflationary expansion and a transition from inflationary back to noninflationary expansion.[129]

Is This Any Way to Run a Science Program?

We will not in this book be answering the big cosmological questions of the age, shape, size, and future of the universe. Current data from astronomical observations are helping narrow down the possibilities, but there are today too many "moving parts"—too many competing possible ways to put the whole picture together—for cosmologists to have reached a consensus view.

It seems legitimate to ask: Is this any way to run a science program? Is this extraordinary range and variety of physicists' research focuses and approaches desirable? Is it a good or a bad sign that we are offered so many different explanations for dark matter, dark energy, inflation?

Physics' world of research and publishing is bit of "democracy in action," and we're reminded of Winston Churchill's observation that democracy is the worst form of government, except all others. We're also reminded of how it's best not to observe the making of either sausage or legislation—and apparently wild scientific theories also.

Mathematician and physicist Adrian Kent comments: "As the twentieth century draws to a close, theoretical physics is in a situation that, at least in recent history, is most unusual: there is no generally accepted authority. Each research program has very widely respected leaders, but every program is controversial . . . So to speak, some impressively large and well organised expeditionary parties have been formed and are faithfully heading towards imagined destinations; other smaller and less cohesive bands of physicists are heading in quite different directions. But we really are all in the dark. Possibly none of us will get anywhere much until the next fortuitous break in the clouds."[130]

Kent's essay, "Night Thoughts of a Quantum Physicist," focuses on physicists' research on topics covered in this book—new physics, consciousness, quantum physics' "measurement problem"—and he credits cosmology as physics' unique success story of the past fifteen years. Perhaps these successes are those that derive from the "fortuitous break in the clouds" of recent astronomical observations, since it's hard to argue that the theorizing on some of the other cosmological topics discussed in this chapter shows an organized direction for the expeditionary parties' headings.

In the democracy of the world's physicists' research-and-publish structure, the pattern seems to be massive shifts in attention toward certain theories that seize physicists' imaginations due to the robustness of their possibilities. From an earlier chapter, string theory is an example of this; from this chapter, it's inflation. As the theory is subjected to more and more scrutiny, we find that we need new special rules of action, or that new special particles or fields need to be hypothesized, to make the theory match all of the known observed facts.

For example, some note that inflation theory leads to creation of an otherwise unneeded "inflaton" particle and inflaton field. They note that because of this "tailor-made inflaton field with no other purpose," it is "necessary to concoct elaborate schemes to get rid of the particles so that they are not observed in the current epoch."[131] And others note inflation theory's need for both a "dilaton" field and an "inflaton" field.[132]

Still others have noted that many of the past two decades' contributions to how inflation theory comes together "were not a stickler for the fundamental physics, such as the standard model or grand unified theories in particle physics."[133]

And there is the continuing concern about the "very 'special' initial conditions" and "'fine tuning' issues" that are raised by various aspects of inflation theory.[134]

Smolin, whose *Three Roads to Quantum Gravity* we discussed in Chapter 12, quantifies the extent of how special inflation's assumed conditions are as having a probability somewhere between one chance in a billion trillion (10^{21}) and one chance in a billion trillion trillion trillion trillion trillion trillion (10^{81}).[135] Your two bucks would be better spent at the track, even at the shorter of these odds.

In effect, some consider the theory of inflation "more on the level of a paradigm than of a theory of the early universe".[136] In other words, inflation provides a highly useful and robust model that incorporates many observed effects. But is it (as is perhaps string theory) becoming a bit of a Christmas tree in terms of the embellishments needed to have the theory reproduce and even predict what physicists and astronomers are able to observe? And can the phenomena that inflation was developed to explain—flatness, horizon, monopole, big bang—be explained without the Christmas tree?

***

We are examining highly accurate theories of physics—inflation, string theory—that are also theories expressed with great economy, each with just a single centralizing concept at its heart. These are formidable theories for anyone to challenge, yet some do.

Let's continue to explore phenomena of new physics, continue to firm up our base of knowledge about observations that some find disturbingly discordant with both physics' standard models and its mainstream enhancements.

What other phenomena lead some physicists toward radical theories of new physics?

15.
Entanglement

Entanglement is a feature of quantum physics according to which particles relate superluminally, faster than the speed of light. Albert Einstein never fully accepted this implication of quantum physics' uncertainty principle, and he referred to entanglement as "spooky action at a distance."

Einstein and two colleagues put forth what they thought was a paradox so absurd that they intended that the uncertainty principle—and its logical conclusion of entanglement—would be laughed out of the world of science. But entanglement has turned out to be fact, even though it "entangles"—instantaneously, and therefore faster than the speed of light—characteristics of far-distant particles.

Today, experimental physicists demonstrate the phenomenon of entanglement and consider how to take advantage of the phenomenon, for example for quantum computers. And theoretical physicists consider the implications of entanglement for the unification of general relativity with quantum physics, for the structure of space and time, and for theories of consciousness.

***

The EPR Paradox

Albert Einstein was fond of "gedanken experiments"—thought experiments, "experiments" conducted only by thinking about them. In 1935, he and two Princeton colleagues (Boris Podolosky and Nathan Rosen) put forth what has become known as the "EPR" thought experiment, by which they intended to disprove a central element of quantum physics.

The element of quantum physics that "EPR" intended to disprove—in fact, to show as absurd—is quantum physics' concept of the uncertainty principle. The uncertainty principle says that there are limits to how precisely we can measure the combination of a particle's momentum and the same particle's position. At an extreme, if for example we very precisely measure a particle's momentum, then this near-perfect knowledge about the particle's momentum will be offset by essentially no knowledge at all about that particle's position.

The EPR experiment (also called the "EPR paradox") proceeds as follows.

Suppose we measure the momentum of a pair of interacting particles right when they meet, just before they bounce off each other and fly apart. Later on, we obtain a precise measurement of the momentum of one of these particles, which is now far distant from the second particle. Since the experiment assumes that this is an isolated system, we can assume that momentum is exactly conserved from the point of our initial measurement forward, so in measuring the momentum of the first particle we can also conclude what the momentum is of the second particle, just by subtracting from our original measurement of the total momentum.

What havoc have we wreaked just by doing this?

By precisely measuring the momentum of particle number one, we have instantaneously (that is, faster than the speed of light) obtained information about particle number two: its momentum is now determined precisely. We also know that, according to the Heisenberg uncertainty principle, we have now lost our ability to obtain any information at all about the position of particle number one: this position is fully uncertain, as a consequence of the precise certainty of the measurement of particle number one's momentum.

But the really astounding conclusion is that the position of particle number two has become fully uncertain. Measuring particle one's momentum has instantaneously affected our ability to measure particle number two's position. By precisely determining particle one's momentum, we have made uncertain particle two's position. There would be no way for an observer at particle two to measure particle two's position.

Einstein could never fully accept this and for the rest of his life argued that the inherent features of these two distant particles are in fact definite—it's only that their measurements are unknown. The side for which Einstein argued has been labeled classical ignorance, the ignorance of not being able to measure a feature that does truly exist with definite magnitude, as opposed to quantum ignorance, the idea that this feature cannot be measured because its magnitude is truly indefinite.

Einstein believed that the ignorance was only classical. He would not even accept the argument of Niels Bohr and other quantum theorists that this quantum ignorance is actually an extension of Einstein's own concept of observer-created reality which Einstein promulgated with his special theory of relativity's concept that space and time depend on each observer's state of motion.

The mainstream view among physicists has been that—odd as it sounds—entanglement is a reality. But it took fifty years before a convincing experiment was performed to confirm this reality.[137]

Experimental Demonstration of Entanglement

In 1986, Alain Aspect published the results of his experiments conducted in Paris, demonstrating the phenomenon of entanglement. The EPR team had not expected experimental testing to be possible for their thought experiment, but experimental physicists had been trying for years to set up such an experiment.

Aspect's experiment is not based on the same application of the Heisenberg uncertainty principle—its implications for the measurement of momentum and position—that was discussed in the EPR paradox itself. Instead, Aspect based his experiment on the measurement of the polarization (which can be thought of as a subatomic trait that is either "up" or "down") of two photons that are produced simultaneously by stimulating an atom. A photon's polarization is determined in part by the angle at which the polarization is measured. Aspect was looking for evidence of entanglement—spooky action at a distance—in the quantum phenomenon of polarization.

Under the Aspect experiment, the analog to the EPR paradox relates to what are the consequences for a particle pair's second particle of determining the polarization of the particle pair's first particle. Since the creation of the pair of photons could not have changed the aggregate polarization of our atom/photon/photon system, it must be the case that the aggregate polarization of the two photons is zero—that is, one photon has "up" polarization, the other has "down." In analogy to the EPR thought experiment, Aspect and team fired pairs of photons, each pair having been simultaneously created from the same atom, toward polarization-sensitive filters.

The classical interpretation has it that, at the point of the photons' creation, one specific photon has been created with up polarization and the other with down. It may be difficult to observe this direction of polarization, but according to classical physics it is already a characteristic of each photon, even before the polarization is observed. Under the classical interpretation, no adjustments of the measuring system will affect the photons' polarizations, which are determined at photon creation.

By contrast, the quantum interpretaion has it that neither photon has a specific polarization before it is observed; both photons are—oddly—in both states, and this will be reduced to just a single state only upon observation.

So this is what Aspect is testing—the classical vs. the quantum picture of the world. The quantum interpretation will require that a measurement of one photon's polarization will instantaneously set the second photon's polarization to the opposite value, no matter where in the universe that second photon is.

Compared to earlier experimenters' versions of this experiment, Aspect enhanced his by adding an additional level of randomness: the selection of the measuring filter used for each photon was not made until the photon was in flight, after the photon was created. This avoids an argument of some kind of advance "knowledge" by the photon of which filter will be measuring it.

Because a photon's polarization is determined in part by the angle at which the polarization is measured, photon one will be influenced to show a direction of polarization based on the angle at which its filter is set. If entanglement—quantum action at a distance great enough that information can be transmitted only superluminally—is correct, then photon two will be influenced in the opposite direction, varying as photon one's filter angle varies.

As each pair passed through the filters, Aspect measured the polarization of one photon, and his experimental findings showed that this measurement of the first photon affected the polarization of the second photon in exactly the way that quantum physics would predict, based on instantaneous coordination of the first photon's polarization with the second's.

Aspect's experiments are considered to be groundbreaking in their confirmation of entanglement's spooky action at a distance.[138]

Nonlocality

Entanglement is an example of quantum physics' nonlocal phenomena, according to which quantum entities such as photons or electrons can be affected by circumstances beyond their own local environments. Local means within a range that can be traversed no faster than the speed of light; instantaneous or any other form of superluminal comunication is therefore nonlocal.

The "double-slit" experiment is a familiar demonstration of quantum nonlocality. It has been carried out on many occasions in many forms, and demonstrates physical instantaneity that cannot be explained classically.

The great physicist Richard Feynman has described the double-slit experiment as encapsulating the "central mystery" and the "heart" of quantum physics. Fortunately for us, Feynman goes on to describe this experiment as containing "the only mystery . . . the basic peculiarities of all" quantum physics.[139] So imagine how much is learned by understanding the double-slit experiment!

This experiment's original purpose, in the nineteenth century, was to demonstrate the wave nature of light. If a stream of light is sent through a single slit, it will create a pattern expected of a stream of particles—intense light directly behind the slit, and less and less intensity further from the slit. But if the light is sent through two slits, what's created is not two sets of the one-slit pattern. Rather, what's created is an interference pattern—alternating stripes of light and dark areas, indicating that light's waves are augmenting or canceling out each other depending on how their peaks and troughs coincide.

Quantum physics asks: How does light "know" there are two slits rather than one? How can a light particle passing through a slit "know" whether a second slit is open or closed?

The demonstration of the nonlocal nature of the quantum world has variations that make it even stranger. For example, the light transmission can be dimmed down so only one photon (light particle) at a time is transmitted toward the two slits. Even when just single photons are passing through the slits, the usual interference-derived pattern will be seen, building up one dot at a time, as if each photon "knows" where past photons have landed or future photons will land. But if we set up a detector at each slit to observe (before the photons strike the screen) which of the two slits each photon passes through, the interference-type pattern will not be created; instead, we'll see two sets of the one-slit pattern, indicating that we migrated the quantum event to a classical event as a result of the observation at the slits, resulting in the build-up of the pattern at the screen conforming to classical not quantum rules.[140]

This is, as Feynman previewed for us, the heart and the central mystery of quantum physics. There is a quantum phenomenon of nonlocality, and entangled particles exemplify this phenomenon. Entanglement is a fascinating phenomenon with rich potential applications and implications.

Applications of Entanglement

Quantum computation—taking advantage of entangled quantum particles to permit computers to operate at superluminal speed—is being actively developed. Physicist Roger Penrose makes the point that the vast majority of quantum states are entangled, opening the door for the potential of massively parallel quantum computation. Penrose goes on to consider more closely how such quantum computation might proceed, considering in particular the fact that quantum entanglement permits only quantum information, not classical information, to be transmitted.

Suppose Alice wishes to send Bob (Penrose's example) information by taking advantage of quantum entanglement. The information will have to be sent bit by bit, using the capacity of the EPR-entangled particles that Alice and Bob each possess. For example, Alice and Bob's quantum teleportation could be taking advantage of the direction of spin (still Penrose's example) of pairs of entangled particles. Alice builds up the bits of the message that she wants to send, based on the direction of spin of her particles, and instantaneously Bob can read Alice's message, based on the opposite spins (the total spin is 0 for the original EPR-pair source) of his half of the EPR pairs.

Penrose traces the process as (1) Alice's quantum information going backward in time to the original EPR-pair source, then (2) forward in time to Bob's half of the EPR-pair, resulting in (3) instantaneous transmission of a bit of information.[141]

Others point out a different aspect of quantum physics which quantum computation might take advantage of: the set of parallel quantum possibilities for how the wave function might collapse. Thus quantum computation could take advantage of quantum physics by exploring all possible quantum options at once instead of serially.[142]

Caltech physicist John Preskill discusses possible future directions for quantum computation. He discusses the implications for precision measurement, quantum error correction, and—where "Big Science will meet quantum measurement"—the use of quantum information theory to aid in the Caltech Laser Interferometer Gravitational-Wave Observatory's quest to detect gravitational waves and other weak forces.[143]

Also intriguing is Preskill's commentary on many-body entanglement—that is, not just entangled quantum pairs. Here Preskill makes the point that the physics and mathematics of many-body entanglement is so complex that we'll be able to take advantage of the implications of many-body entanglement only if we can build a quantum simulator to learn by experimentation rather than by theoretical modeling. "Good news for experimenters", he says.[144]

Quantum computers are just one element of the rich lode of potential applications being hypothesized for quantum entanglement. The vocabulary alone fascinates: tachyonic propagators, transluminal matter, traversable wormholes, dark matter remnants of imaginary time.[145] We cannot imagine today where these applications may take us.

***

For theoretical physicists, too, entanglement has implications.

Up until very recently, physicists generally agreed that the core "unsolved problems of theoretical physics pertained only to the domain of quantum gravity, that is the conjunction of quantum [physics] and general relativity."[146] However, some physicists now wonder if a transformation in theoretical physics is evolving out of current research on entanglement, along with research on macroscopic quantum phenomena and other phenomena of new physics that we discuss in Part Three.

When all of these additional research areas are looked at together, some physicists draw conclusions that extend well beyond today's formulation of the questions, toward radical rethinking of the nature of space and time, as well as toward a critical role in physics for consciousness and the mind.

16.
Entropy and Information

Entropy is the opposite of orderly process, of focused, productive energy. Entropy is dissipation, randomness, not acting in concert.

Throughout our lives, we're constantly fighting to keep our entropy low—that is, to keep our self-order and self-organization high. This is true for our biology as well as for our minds.

Biologically, we're constantly taking in low-entropy energy and shedding high-entropy energy. Our intake—the food we eat and the air we breathe—is more ordered (lower entropy) than our outflow, especially the heat we release, but also our biological excretions. Heat is the highest entropy (least ordered) form of energy, so we're decreasing our entropy every time we take in a highly ordered piece of steak and release highly random heat energy into the environment.

Plants, too, stay alive through their intake of nutrients, water, and atmospheric gases, and their release of simpler oxygen gas.

And intellectually, as we increase our understandings through organizing information, we are lowering the entropy of our local universe, creating a more ordered, less random environment.

But we're swimming upstream with these life functions that fight to keep order.

You probably remember various conservation laws of physics, such as the law of conservation of energy. But entropy is not conserved; it increases. A closed system's entropy increases with the passage of time. This is a basic principle of thermodynamics. And this is also virtually unique as an indicator of the arrow of time, a rare instance of physics that marks the forward direction of time.

***

The Arrow of Time

Entropy is defined as the extent of disorder.

This seems like a funny thing to care about, except for one basic concern of physics. This basic concern is the arrow of time—why time proceeds in the direction from yesterday to today to tomorrow, and only in this direction.

A basic principle of thermodynamics is that entropy in a closed system increases with time. Things fall apart, become less ordered. This is a rare instance in physicis in which time has an arrow.

In virtually all of the rest of physics, interactions can take place in either direction of time. Suppose we take a moving picture of two particles A and B crashing together, annihilating each other, then forming new particles C and D, which move outward in different directions. Then we show this movie in reverse to an assembly of physicists. No one would blink an eye or wonder what's going on. The movie played in reverse would look perfectly natural: particles C and D come together, annihilate each other, and form particles A and B, which move outward in different directions.

The laws of physics for particle interactions do not have an arrow of time, and there is no way for the assembled physicists to know whether they're watching the movie in the opposite direction from what actually took place.

This is what makes entropy such a fascinating measure for physicists: it provides a direction for the arrow of time.

University of Oxford physicist Stephen Hawking, in *A Brief History of Time*, discusses three senses in which we experience an arrow of time: the psychological time that we experience in our minds as proceeding from the past toward the future, the future-directed time orientation of increasing entropy, and the universe-expanding direction—as time goes on, the universe expands.

Hawking describes how, earlier in his career, he thought that time's arrow would reverse if the universe stopped expanding and began contracting. In other words, we would begin to experience the future as taking place before the past. However, Hawking's current position is that there would not be a reversal in the direction of the arrow of time if the universe begins to contract, which leaves entropy as the unique physical phenomenon aligned with our psychological experience of time.

But what would this possibly mean, to experience the future before the past?

Roger Penrose discusses what cause and effect would feel like if time's arrow were reversed. We're used to causes taking place before effects, but this ordering would be reversed if time's arrow were reversed. What would happen is that we would develop a psychological sense that an observed effect must naturally be followed by what caused that effect. If we saw a coffee cup lying broken on the floor, we would expect to soon see what caused this to happen—for example, someone knocking the cup over. This would not surprise us, in a world in which time's arrow was reversed. In fact, we would expect life to proceed in the order that first we see effects, then soon thereafter we'll see what caused those effects.[147]

The Relentless Increase in Disorder

When a gas is let into a room, initially the gas is concentrated at the point of intake. But soon the gas disperses throughout the whole room. This is entropy in action: the gas moves from a narrow specific location within part of the room to a uniformly dispersed spreading throughout the entire room. Entropy has increased, because at intake we locate the gas specifically, not randomly. After dispersion, we can speak of a general level of concentration of the gas throughout the room, and we do not need to specify additional information to describe the status of the gas. The earlier low-entropy state required a level of specific description—where within the room there were various concentrations of gas—that is not needed to describe the later high-entropy state.

When we pour cream into coffee, initially this is low entropy: it's not disordered or random; we need to include a description of where the cream is and where the coffee is in order to describe this state. But after the cream and coffee are mixed, we have a higher-entropy state: we can describe the coffee/cream mixture within the coffee cup just by describing the overall mixture, without a need to add a description noting parts of the contents being coffee here and cream there.

Other common examples also demonstrate this basic principle of thermodynamics that entropy increases with time. When we let a blast of cold winter air into a heated room, the natural tendency is for the air to mix together and the temperatures to equalize throughout the room. Physicist Michio Kaku even cites the slow rise and rapid fall of civilizations to demonstrate the inevitability of the march of entropy with time. Bleakly, Kaku goes on to sketch the expanding universe's path to absolute zero, to nothingness, to cosmic heat death, unless a rare quantum event pulls the universe along a more hopeful path.[148]

But Roger Penrose asks us not to be glib about entropy always increasing. After all, this would require entropy at the beginning of time—at the big bang—to be very low. Where else would entropy be increasing from? Penrose spends many pages of *The Emperor's New Mind* addressing this question, and his answer takes us to the very special and unlikely traits of the big bang, through quantum gravity, and on to the physics of the mind.

A Rich and Robust Concept

Entropy is a robust tool for understanding theories of physics.

Black holes, discussed in more depth in the next chapter, have proven to be a surprisingly important feature of physics. String theory's success at explaining the mysteries of the entropy of black holes is cited as support for the validity of string theory.[149]

Physicist Julian Barbour notes odd implications of entropy's arrow of time, then uses this to suggest radical rethinking of the role of time within physics: "It may be easier to explain the arrow of time if there is no time."[150]

Information

Entropy is a measure of information content.

Information is measured in bits: 0s and 1s, yes's and no's. In the game Twenty Questions, we get to use twenty bits of information, because we get to ask twenty yes/no questions. On our computers, we can store many megabytes of information (if a byte consists of eight bits, a megabyte is eight million bits).

When a gas is first let into a room, it requires an additional level of information to describe the room than it will take after the gas is fully dispersed. The earlier, more ordered state—the lower entropy state—requires more information to describe it. Lower entropy means higher information content.

The equivalence between (low) entropy and (high) information can be taken just a bit further, to create what must be among the most surprising relationships anywhere: equating information and temperature. Physicist Ludwig Boltzmann has developed a formula connecting the entropy of a system to the number of accessible microstates, which is how he quantifies information. From this he derives the heat price of information: to extract one bit of data costs one ten-thousandth of a trillionth of a centigrade degree (10^{-16} °C).[151]

One way to define entropy is the amount of information that is hidden behind a macroscopic description. For example, when a gas has been fully dispersed within a room, we are able to describe the state just by describing the density of the gas within the room. Behind this description, we have hidden large quantities of information about where each individual molecule of the gas is located. In the high-entropy situation, we can describe in simple terms the general situation, which means we don't need to say anything about—we can leave hidden—the specific information about each molecule. This contrasts with the earlier low-entropy situation, when we have just begun to inject the gas into the room. To describe this earlier, low-entropy state, we do need to provide additional information about the location of specific gas molecules. So there is less hidden information, that is, less entropy.

Some bring information to the center of quantum theory. The reduction of the wave function is the process of gaining information by the transition from potential information—a description of possibilities of the future—to actual information after a measurement is performed. Facts are classical and ireversible: quantum possibilities have been reduced to a single classical event.[152]

Information has been put forth as a centralizing concept for all of physics. New physics places the transmission of information from the past to the future as a process at the heart of the structure of spacetime.[153]

This leads to the connection with this book's subject, new physics and the mind. Manoj Samal, of India's S. N. Bose National Centre for Basic Sciences, puts forth in "Speculations on a Unified Theory of Matter and Mind" the argument that information—more than matter, more than energy—is the most fundamental entity of the universe and one that is the central concept for the unification of matter and mind.

New physics' theories of quantum gravity and theories of the nature of space and time are theories of the organization of information.

**

It is perhaps ironic that we seem to have made more progress in our understanding of information as a concept within the hard science of physics than we have in rationalizing the concept of information as a scientific tool for the study of consciousness and the mind.

17.
Black Holes

Science fiction has prepared us well for our understanding of black holes.

Black holes are large quantities of mass concentrated in very small spaces. A frequently cited measurement of this is that a teaspoonful of black hole would weigh as much as Mount Everest. Nothing, not even light, can escape a concentration of mass this great, a mass that will never be illuminated, a black hole.

Black holes are structurally very simple, a point that physicists sometimes summarize as "black holes have no hair." Black holes have little in the way of features that allow us to distinguish one black hole from another. The primary feature of a black hole is its mass, which determines how small the black hole will be.

The existence of these extraordinarily dense concentrations of matter is universally accepted by theoretical and experimental physicts, by particle physicists and astrophysicists. And the nature of black holes provides an important window to many aspects of modern physics.

***

Black Holes in the Universe

Cosmologists believe that black holes are fairly common, formed at the end of the life cycle of many stars. There are estimated to be one hundred million black holes within our galaxy, the Milky Way, alone.[154] Astronomical evidence suggests that a black hole with 2.6 million times the mass of the sun is at the center of the Milky Way, that supermassive black holes which power quasars can be ten thousand times larger, and that little black holes proliferate throughout the universe.[155]

Our sun is expected to use up all of its fuel over billions of years. As the sun's fuel is used up, the decreasing nuclear fusion reduces the sun's push outward against its inward-pulling force of gravity, and gravity will crush the sun inward, until the sun becomes a white dwarf when the inward and outward forces are once again in balance. This is projected as the future of our sun: a white dwarf that will then slowly die out.

But stars that are a few times larger than our sun will not die with a whimper—they'll have enough energy for one last kick, and a big one. These stars will explode as a supernova.

And stars that are larger still—ten to fifty times the size of our sun—will have yet a different future. Gravity is too strong for these stars to burst out as a supernova. Instead, gravity will continue to squeeze these larger white dwarfs. They will become black holes. The black hole is the region around the collapsed star in which the gravitational force is so great that nothing—not even light—can escape.

The outer surface of a black hole is called its horizon—also called its event horizon—and the region behind the black hole's horizon is its hidden region. Nothing that crosses the horizon into the inside of the black hole will ever leave the black hole. The hidden region is hidden to those outside of the black hole's horizon, although the hidden region would be visible to anyone inside the black hole.

The universe is fourteen billion years old. Star formation and extinction has been going on for many years. Throughout the universe, stars are being born and stars are dying. Although black holes still have not been observed astronomically, there are a number of strong candidates for black holes seen in the skies, and most scientists believe in the existence of black holes.[156]

Wormholes

Topological models show that a black hole does not necessarily narrow down completely into a dead-end singularity, but instead may open up to another black hole or even another universe. The connecting tunnel is a wormhole.

Wormholes have never been observed, but the possibility of their existence comes up repeatedly as mathematical models show that they could exist. Their existence would raise exciting possibilities about connecting regions of space that would otherwise be far distant.[157]

Entropy and the Holographic Principle

The area of a black hole's horizon is a measure of the black hole's entropy. This has a few interpretations. For one, based on the principle that entropy represents hidden information, the area of a black hole's horizon measures the amount of information hidden within the black hole. Our observation that what falls into a black hole is irreversibly irretrievable is analogous to the established thermodynamic principle that entropy increases irreversibly with time: what's hidden behind the black hole can only increase in size.

For some physicists, black hole thermodynamics leads to a possible understanding of quantum gravity. Black hole thermodynamics is one of Smolin's "three roads to quantum gravity," along with string theory and (his own specialty) loop quantum gravity. Black holes act as microscopes that allow us to observe the physics that operates on the Planck scale, the smallest scale of physics and the scale at which gravity must operate as a quantum phenomenon. Smolin's unification of the "three roads" is via the holographic principle, the principle that the whole is reflected in each part, and that the quantum information encoded in a volume of space is captured in the surface that bounds this volume. For a black hole, the area of its horizon holds all of the information hidden in its volume.[158]

What's Going On With . . . Black Holes

The distortion of spacetime by black holes provides rich source material for science fiction: black holes help us make oddly fast journeys through vast distances of space, and they allow us to travel backward or forward in time. Scientists debate, and are skeptical about, the practicalities of these fictional journeys. But scientists also find black holes to be a concept of extraordinary value in developing theories about many aspects of modern physics.

The scientific research literature on black holes is enormous in size and scope. Black holes are being studied by physicists throughout the world, and these studies are being applied to string theory, studies of extra dimensions, cosmology, theories about the nature of time, matter, and space, and other fields of physics across the board.

Here's a sampling of research on black holes. This sampling is being performed at over forty universities and research institutions throughout North and South America, Europe, and Asia.

Black holes indicate such extreme distortion of spacetime that holes appear in the spacetime fabric. Some theorize that through these holes are parallel universes—key to understanding relativity, the quantum world, and physics' theory of everything.[159]

Stephen Hawking has theorized extensively on black holes and their implications for cosmology, the development of the universe. A beginning point of his conclusions is the observation that black holes represent a singularity in spacetime. From this simple beginning, Hawking draws broad conclusions about another important singularity of the universe, the big bang. The applications are far-reaching, including conclusions about quantum cosmology and the *wave function of the universe.*

Hawking is also known for explaining what is now known as Hawking radiation. This appears to be the release of particles from a black hole, which seems contradictory to the insatiable appetite that is at the heart of a black hole's nature. Hawking explains this instead as the escape of a nearby particle which is just outside of the black hole's horizon when its entangled particle passes through the black hole's horizon never to be heard from again.[160]

Experimental physicists consider possibilities for observation of black holes both through high-energy colliders and, in outer space, through large observatories, perhaps providing our first opportunity for experimental study of black holes by observing those that have been created by high-energy cosmic rays.[161]

Some physicists look at possibilities for black holes having been created at the big bang (that is, "primordially")—created in a higher dimensional universe.[162] They look at primordial black holes as candidates for dark matter,[163] and explore quantum black holes (not the simpler classical black holes) and consider how quantum fluctuations may have produced primordial black holes.[164]

Black holes have frequently been considered in the context of extra dimensions. Physicists discuss "windows to extra dimensions" near black holes,[165] and demonstrate that the topology of black holes can be equated to the topology of compactified and noncompactified extradimensional spaces.[166] Others propose how theoretical models of black holes, as well as black holes created in high-energy colliders, can help us explore the geometry of extradimensional spacetime.[167] Physicists find "black diamonds at brane junctions"—black holes shaped as diamonds and polyhedrons at the intersections of our $(3 + 1)$-dimensional membrane ("brane") with other branes of the extradimensional bulk.[168]

Other physicists are studying the evaporation of black holes,[169] quantum entanglement of particles behind and outside of the black hole's horizon,[170] and the black hole horizon itself, which separates the inside from the outside of the black hole and must have some thickness, even if only of Planck length.[171] Many physicists study black holes' related topological phenomenon, the white hole: black holes are viewed as windows to higher dimensions, and white holes as windows from higher dimensions.[172]

***

There is an amazing amount of thinking going on about a phenomenon—black holes—that has never actually been observed. This is part of physics' great tradition of progress through thought experiment.

We will see again, in Part Four's Ten Radical Theories of New Physics, a number of the concepts introduced in this chapter—holographic capture of a spacetime volume's information, entanglement between the inside and the outside of the black hole, the black hole horizon itself, black diamonds at brane junctions, wormholes, parallel universes.

Black holes provoke much thought about new physics and the mind.

18.
Imaginary Time and Multiple Histories

University of Oxford physicist Stephen Hawking specializes in cosmology—the science of the origin, current state, and future of the universe. Hawking's concepts of cosmology, and his concepts of what must constitute a Theory of Everything, involve the idea that the universe has multiple histories, not just the history that we participate in, and also involve the idea of imaginary time.

***

The Sum over Histories

One important way to understand quantum physics is the sum-over-histories approach, originally developed by physicist Richard Feynman. Feynman won the 1965 Nobel Prize for physics for his work on quantum electrodynamics, which brought classical physics' electromagnetism into the world of quantum physics.

Sum over histories can be very hard to believe in. It proposes as a scientific principle that we participate in only one of the universe's multiple histories.

According to Feynman's sum over histories—which also goes by the names sum over paths and path integrals—a particle travels from point A to point B by all possible paths. Not just the direct route that we seem to expect. But also by all possible indirect routes, including zigzags, curves, via the moon, and via Mars.

This conceptualization creates a framework for understanding quantum physics' probabalistic approach to the physical world. Quantum physics does not give us precise answers to where a particle is and how fast it is moving. Instead, quantum physics gives us a probability distribution: what is likely, and what is unlikely but nevertheless possible.

At the macroscopic level of our everyday lives, all matter and activity are predictable by classical physics because the unlikely possibilities of the subatomic world cancel each other out and are impossibly unlikely. But at the level of the atomic world, observation collapses the wave of possibilities into single observed points. This collapse of possibilities proceeds according to a predictable range of probabilities, but the specific observed outcome is unpredictable.

What happens to the whole range of possibilities that our observed world did not collapse into?

The sum over histories answer is that they still exist, but not in the observer's universe. They exist as another history of the universe. The universe has multiple histories, in only one of which is a particular observer residing.

Imaginary Time

Feynman's sum over histories is not a straightforward summation using real numbers. Instead, it requires weightings by imaginary numbers. In particular, the summation takes place in imaginary time, not the real time you and I experience in our daily lives.

For mathematicians, imaginary numbers are no more or less real than real numbers. I will probably not convince you of this if you don't already believe it. But the mathematician's perspective is that, when you delve deeply into real numbers, you realize that they're not that real either—they're refined formal representations of deep numerical concepts. And this is what, to mathematicians, imaginary numbers are also.

The power of imaginary time in particular is that it brings time to an equal footing with the three dimensions of space. It makes calculations in the four dimensions of spacetime a lot simpler when the time dimension is measured in units of imaginary time.

And in Hawking's world of quantum cosmology, the universe is simpler to explain and to understand when the fourth dimension is imaginary time.

The Universe Is Finite but without Boundaries

Normally, you'd expect something to be finite with boundaries. Or else it can be infinite and without boundaries. But you'd be wrong if you thought you're forced into just these two choices.

A familiar example is the surface of Earth. Earth's surface is finite—

it does not go on forever. But there are no boundaries to the surface of the earth.

Hawking's quantum cosmology contemplates that the universe is finite but without boundaries. However, this is true only in imaginary time. In real time, the universe had a beginning—a discontinuity, or a singularity—at the big bang. And it will have an end, either by recollapsing in a big crunch, or expanding forever into nothingness.

But the universe in imaginary time has no singularity because it has no boundary. It has no point at which the laws of physics break down. In this sense, it leads Hawking to wonder whether imaginary time actually provides a more fundamental understanding of the concept of time than real time provides.[173]

***

Physicist Michio Kaku writes at length about colliding universes, vividly bringing to life Hawking's applications of quantum physics to the cosmology of the universe. As Kaku describes it, Hawking's cosmology creates parallel baby universes, each connected to each other by wormholes.[174]

This is a vision of physics also included within some of the models of new physicists who look to include consciousness and the mind within their understanding of a theory of everything. We'll see more of this picture in Part Four of this book.

19.
Tunneling

We've all seen the bell curve of the normal probability distribution. There are also probability distributions besides the normal curve, and many of these probability distribution curves share the trait of bunched up (higher) probabilities for some central outcomes, and remote probabilities for extreme outcomes.

Quantum physics has changed our notion of the localization of elementary particles from a single specific location to a probabilistic distribution of various locations where a particle could possibly be.

This means that some very unlikely locations can have small, but not zero, probabilities of occurring. For example, there can be probabilities—very unlikely, small probabilities—of particles tunneling through barriers that would typically (classically) be considered impenetrable.

The probability of macroscopic "tunneling," such as our house being transported to Mars, is just too remote to occur—at least in our lifetimes, or the lifetime of the planet or even the universe. This is because we'd have to have simultaneous tunneling of all of our house's elementary particles to the same unlikely location at the same time.

But the phenomenon of tunneling is entirely well established at the level of elementary particles, where experiments match the predictions of quantum physics with great precision.

***

An example of quantum tunneling is the radioactive alpha decay of heavy nuclei such as the uranium nucleus, in which two protons and two neutrons held tightly together are emitted as an alpha particle.

According to calculations of classical physics, these alpha particles should not have enough outward-pushing electromagnetic force to overcome the strong force holding them within the nucleus. But quantum physics' uncertainty principle correctly predicts that some alpha particle outliers will be located just far enough away from the nucleus's center—even if just for an instant—to escape from the strong force's control. This is an example of quantum tunneling: the small and brief variation in the location of an alpha particle permits it to tunnel through the barrier that the strong force normally creates.

Another example of quantum tunneling occurs as part of the process of nuclear fusion. Here, adjacent protons exhibit tunneling behavior, which permits them to tunnel into each other's outer boundary, crossing through the electromagnetic barrier that repels two protons. Again, classical physics calculates that protons will not penetrate this barrier. But with quantum physics' creation of more extreme—although low-probability—possibilities, it is in the outlying, not the average, situation that the protons tunnel through the barrier between them.

What tunneling phenomena have in common is that a classical understanding, without quantum physics' uncertainty principle, would not permit this extreme behavior, this crossing of an energy barrier that is considered impenetrable classically. But in actual practice, the extreme behavior takes place with the precise frequency predicted by quantum physics' uncertainty principle, a central theory of quantum probability.[175]

What's Going On With . . . Tunneling

Throughout the world, theoretical physicists are studying the implications of tunneling, and experimental physicists are exploring applications.

IBM scientists in 1981 successfully took advantage of quantum tunneling to develop the scanning tunneling microscope. Today, scanning tunneling microscopes permit individual atoms to be observed and manipulated.[176] Physicists Gerd Binnig and Heinrich Rohrer were winners of the 1986 Nobel Prize for physics for their design of the scanning tunneling microscope.

Other physicists look for applications of tunneling to quantum computation. Frequently, this involves taking advantage of the tunneling behavior of neighboring quantum dots, which are extremely small conducting devices containing up to several thousand electrons. Experimental physicists perform detailed studies of the degree of tunneling conductance through quantum dot systems. Strangely, physicists observe different patterns depending on whether an odd or an even number of electrons occupy the quantum dot.[177]

Not to be outdone, theoretical physicists propose applications of tunneling at every scale, from subatomic to astronomical. In quantum models of the development of the universe, quantities of energy or mass appear spontaneously then quickly disappear. Quantum tunneling permits the universe to arise from nothing, tunneling through the potential barrier, then continuing to expand to the size of our current universe.[178]

Others, looking to the distant future, model possibilities for our universe eventually destabilizing by an ending quantum tunneling, noting that aspects of this work are "only of academic interest to us, since we will not survive the tunneling."[179]

Physicists are investigating the possibility of the production of black holes from the collision of particles through quantum tunneling processes.[180] And tunneling has been put forward as both an explanation for Hawking radiation, which mysteriously appears to escape from a black hole,[181] and as an alternative ("gravitational tunneling radiation") to Hawking radiation as an explanation for black holes' apparent radiation escape.[182] Physicists are developing a census of which types of particles can tunnel out of the interior of a black hole, depending on whether the black hole is static (no tunneling), non-extremely rotating (only photons tunnel), in extreme rotation (any particle may tunnel), or charged and rotating (still being studied).[183]

Some physicists look at how the mathematics of chaos, complexity, and catastrophe relate to the phenomenon of tunneling.[184] Others look to stretch classical physics to incorporate quantum tunneling, some proposing that classical physics can offer most or even all of the theoretical explanation.[185]

Physicists being who they are, it's inevitable that they'll want to study the traversal time for tunneling. Yes, this is the measurement of how long it takes a tunneling particle to cross the tunnel barrier. It turns out that the conclusion from measuring tunneling time is that tunneling occurs superluminally—faster than the speed of light.

Several theories have been developed to explain the phenomenon of superluminal tunneling traversal time. The explanation needs to include why this superluminal traversal time does not violate special relativity's prohibition against speeds greater than the speed of light. One explanation is based on a distinction between group velocity and front velocity, and there are other explanations involving semiclassical time and the existence of more than one tunneling time. Tunneling time is also modeled using imaginary time and complex time, invoking the concept of instantons—theoretical short-lived quanta of spacetime, which are hypothesized to avoid singularities at which the laws of physics break down.[186]

Some invoke tachyons, hypothetical particles that travel faster than the speed of light, to explain tunneling and its superluminal traversal time. The tunneling behavior of the tachyon is proposed as the possible basis for a hidden variable description of quantum physics.[187]

The topics of other studies of tunneling range from theoretical proposals to explain the origin of time[188] to practical applications of superluminal tunneling time, as a way of speeding up signaling transmission and improving microelectronic devices.[189]

Josephson Junctions

Welsh physicist Brian Josephson studied the join, or junction, between two superconducting materials (materials with no resistance to the flow of current) that are separated by a thin insulating barrier. Even if the materials were ordinary, not superconducting, there would still be a small flow of current across a thin insulating barrier, due to tunneling effects. Josephson realized that the tunneling effects for superconducting materials would mean that the current would flow freely across the barrier, encountering no resistance at all.

Josephson experimented with applying to the junction various levels of current, as well as various strengths of magnetism. He found a whole series of phenomena: superconductivity across the barrier appearing and disappearing, direct current switching to alternating current, sharp switching between allowing current to flow and shutting it off. At age 33, Josephson shared the 1973 Nobel Prize for physics for this work. (Physicists Leo Esaki and Ivar Giaever were co-winners of the 1973 Nobel Prize, for their work on other phenomena of tunneling.) Josephson's work has applications in precise calculations of important constants of physics, including the charge of an electron and Planck's constant. And—if superconductors can become effective at practical rather than only very cold temperatures—his work will have important applications in high-speed low-power computer switching.[190]

Josephson's later work shifted to studies of intelligence, the mind, and extrasensory perception, which was "regarded as somewhat beyond the pale by many of his physicist colleagues."[191] However, Pitkänen—whose theory of topological geometrodynamics is core to the speculations in Part Four of *New Physics and the Mind*—makes important use of Josephson junctions.

Physicists continue their investigations of applications of tunneling to superconductivity.[192] Work continues today on applications of Josephson junctions to quantum information processing and other macroscopic applications of quantum tunneling[193] as well as to quantum logic circuits.[194]

***

Again with tunneling, as with so many other phenomena of science, we see physicists looking for connections. Tunneling and black holes. Tunneling and superconductivity. Tunneling and quantum dots. Tunneling and quantum computation. Tunneling and chaos. Tunneling and the creation of the universe. Tunneling and the end of the universe. Tunneling and the mind.

More pieces to the puzzle.

What will the assembled puzzle look like?

20.
Bose-Einstein Condensates

Another mysterious phenomenon of new physics.

A Bose-Einstein condensate is quantum matter, peculiarly condensed matter. The theoretical basis for the Bose-Einstein condensate was developed early in the twentieth century, but this new state of matter was not experimentally produced until 1995. It has generated significant excitement since then.

Typically, quantum physics does not contemplate that a group of quantum particles, such as bosons (force particles), will all be in the same quantum state, will all act in concert. But now Bose-Einstein condensates have been produced at supercooled temperatures in very small amounts—quantum particles acting in concert at a macroscopic scale.

***

The force-carrying particles bosons are named for the Indian physicist Satyendra Bose, who developed a statistical approach to quantum physics and to black-body radiation. Bose applied this approach to photons, and Albert Einstein generalized this to other particles.[195]

The result was a theoretical prediction of what has come to be called the Bose-Einstein condensate, condensed matter which operates as a giant matter wave. This comes about as a gas is cooled to a temperature very close to absolute zero, at which point the particle characteristics of the gas's atoms are overwhelmed by their wave characteristics. Particles become indistinguishable and form a coherent cloud of atoms, similar to the subatomic alignment that creates lasers and superconductors. This atomically aligned matter—condensed matter—is predicted to permit the frictionless extension of atomic characteristics to macroscopic sizes.

The theoretical development of the Bose-Einstein condensate focused on bosons, force particles whose quantum spin is a whole number. But atoms are also bosonic—even though atoms are composed of fermions, matter particles with spins that are half-integers, not whole integers like bosons—when their total number of electrons, protons, and neutrons is even. In such a case, even atoms can form Bose-Einstein condensates. Physicists working on Bose-Einstein condensates label as "breathtaking" the progress in just the few years since they were first experimentally realized, and they go on to proclaim that this is "more than just a phenomenen of statistical physics: it is a new window into the quantum world."[196]

Bose-Einstein condensates are in a highly entangled state, and applications are being developed in various fields of quantum information, including quantum computers and quantum cryptography. Some emphasize the value for quantum computation when the entanglement is between an atom and a photon, which is the carrier of electromagnetic force. Such entangled atom/photon pairs, created from Bose-Einstein condensates, would be adaptable to quantum computation by taking advantage of the features of each of the two particles: atoms are easy to store at fixed locations, but are not well suited for transmission over long distances, whereas photons are easily transmitted but not easily stored. So the quantum computer structure could use photons to transmit information among stored atoms with which they are entangled.[197]

Applications to teleportation are also under consideration. Here, the procedures under consideration teleport the state of a massive particle, rather than the massive particle itself. But isn't that all that Captain Kirk and his Star Trek team need for their transporter—to teleport the information needed to reconstruct the transported matter at the receiving end? An experimental protocol has been proposed for teleportation of the states of an atomic Bose-Einstein condensate. The protocol is formulated in three stages: entanglement formation, measurement, and receiver operations.[198]

Of direct interest to speculations we will be discussing in our closing chapters is the work by numerous physicists on the design and engineering of quantum devices deploying Bose-Einstein condensates and arrays of Josephson junctions. These devices raise the possibility of entangled webs permitting multifaceted superluminal communication across space and time boundaries that are otherwise impenetrable.[199] Our speculative science of Part Four looks to biological applications of these condensates and junctions. Applications to consciousness are found, including quantum signaling devices permitting communication between the mind and the universe.

Theoretical Interconnections

As rich concepts of physics take shape, physicists frequently look for parallels and interconnections with other established concepts. The connections physicists find between Bose-Einstein condensates and a host of other phenomena of new physics is indicative of how rich and robust is the Bose-Einstein mechanism.

For example, some look to Bose-Einstein condensates to study astronomy's cosmological constant.[200] Others extend the concepts of Bose-Einstein condensates to gravitational systems, offering an alternative to black holes as the solution to the endpoint of a star's gravitational collapse.[201] Still others note that Bose-Einstein condensates mimic phenomena of general relativity and therefore offer a promising system for gravitational measurements.[202]

Theorists find a hidden topological order in certain Bose-Einstein condensates[203] and note similarities between Bose-Einstein condensate systems and extradimensional models of spacetime in which only gravity propagates into the bulk of the extra dimensions.[204] Others look to take advantage of Bose-Einstein condensates' coherent motion under the force of gravity in order to test theories of the shape and size of extra dimensions.[205]

Gravitational Antennas and Transducers

Some see Bose-Einstein condensates as phenomena of quantum gravity, at the interface between quantum physics and general relativity. As macroscopically coherent quantum fluids, Bose-Einstein condensates are macroscopic quantum gravitational antennas and transducers. This means that Bose-Einstein condensates can serve both as receivers and sources of gravitational waves: they convert gravitational radiation into electromagnetic radiation (quantum gravity antennas), and electromagnetic radiation into gravitational radiation (quantum gravity transducers).[206]

***

The 2001 Nobel Prize for physics was awarded to three physicists (Eric Cornell, Wolfgang Ketterle, and Carl Wieman) for their experimental and theoretical efforts on Bose-Einstein condensation.

We discussed in Chapter 11 the work of Zohar and others, including

Bohm's development of the Implicate Order, marking perhaps the dawn of new physics and the mind. Zohar links our sense of "I" to the operation of Bose-Einstein condensates within our neurological and other biological systems. Microtubules within the nervous system, which are a focus of Penrose's and others' physics of the mind, exhibit traits of Bose-Einstein condensation. And Marshall and Zohar note the controversial speculations of physicist Herbert Fröhlich that Bose-Einstein condensation is characteristic of life itself, maintaining quantum coherence organically across the magnitude of living cells.[207]

Bose-Einstein condensates are at the heart of the construction of Pitkänen's topological geometrodynamics, which we will discuss in Part Four. There we will see Bose-Einstein condensates applied in combination with entanglement, Josephson junctions, gravitational antennas, extra dimensions, and macroscopic quantum phenomena.

21.
Chaos and Complexity

In high school physics, we learned a lot about how predictable the physical world is. If we carefully analyze a physical system, and draw a diagram that shows in simple terms the masses and forces that are acting, we can provide answers to all sorts of questions about how high, how far, how fast, for how long a cannonball will travel when shot into the air at a specific angle and with a specific initial velocity. If we're good at this, the answer that we provide is the correct answer. One answer is right and the rest are wrong.

Quantum physics changed this. In quantum physics, some answers aren't knowable with definiteness. They're knowable only probabalistically—what might happen with what ranges of likelihood.

This bothers people. Even Einstein objected: "God does not play dice."

But this quantum uncertainty is a fairly constrained uncertainty, generally understood to operate only at subatomic scales. You won't get an A in high school physics if you say it's uncertain where the cannonball will land.

But there's another kind of uncertainty that's not at all constrained. We see it every day, and we see it at macroscopic levels. The weather. The motion of fluids—gases and liquids. The butterfly that flaps its wings in Brazil and causes a tornado in Texas.

These are complex systems and chaotic systems. And a question that many have is: Do complexity and chaos simply reflect the inadequacy of our models? Or are they physical phenomena in their own rights, adding levels of unknowability to classical understanding?

***

Why is it so hard to model some macroscopic systems, such as weather patterns or the turbulence of air around aircraft?

For complexity theorists, some systems are not merely complicated—they are complex. A complex system is not simply a complicated aggregation of components; a complex system aggregates to a sum that is greater than its parts, an aggregation that cannot be fully explained by its components.

Systems—even those appearing as simple as the rise of smoke from a candle—can be wildly unpredictable. These patterns are studied by proponents of chaos theory, who specialize in systems in which very minor variations in initial conditions result in enormous variations in outcomes. These theorists see the potential for chaos—wild unpredictability—in many phenomena that at first glance seem that they ought to be subject to straightforward rules and formulas of cause and effect.

Chaotic phenomena have been studied since the late 1800s, but interest accelerated in the 1960s and 1970s. Today, theorists find the effects of chaos in astronomy, biology, the rise and fall of animal species, global climate change, even patterns in the stock market.

Chaotic phenomena are nonlinear: they reflect a "hypersensitivity [to] initial conditions."[208] If a chaotic event takes place a second time, either in nature or in a laboratory, with just a small variation in how we start out, the variation in the outcome will expand exponentially rather than linearly as a function of the variation in the initial conditions.

Fractals

Fractals are repeated, embedded patterns—nested shapes—such as are observed in coastlines, plant growth, snowflakes, and other natural phenomena. The term was invented by IBM mathematician Benoit Mandelbrot in the 1960s, from the Latin word for broken stones. Fractals emphasize "self-similarity," the embedded pattern of resemblances between each small part that is embedded in a larger whole.[209]

Complexity theorists consider fractals to be indicative of extra dimensions, infinite dimensions in fact. This formulation derives from fractal geometry's limitless levels of self-imbedding, infinite nesting hierarchy.

Some physicists are exploring the implications of the possibility of a fractal structure to spacetime. Said another way, fractal spacetime is equivalent to spacetime of infinite dimensions. In some formulations, our spacetime world of four dimensions represents an average number of dimensions over an infinite number of possible values.[210] The fractal structure to spacetime is also part of the foundation of topological geometrodynamics, discussed in our closing, speculative chapters.

Physicists look at "fractal signatures in quantum physics and cosmology," and at how a hypothesis of fractal spacetime leads to radical rethinking of space, time, matter, and gravity. These are minority views among physicists—offbeat, not "canonical."[211] But we'll be speculating on these emerging views in our closing chapters.

Detractors and Advocates

Complexity theory detractors wonder how this theory is to be distinguished from a simpler explanation—that complex phenomena are simply too difficult to successfully model with today's science, mathematics, and technology. Thus, they would argue, these phenomena that may look complex or chaotic are simply too complicated to be modeled by today's working algorithms, but this is a far cry from suggesting that these phenomena have any qualitatively different theoretical basis than mainstream scientific explanations.

Complexity and chaos theories have their fervent advocates, however, among prominent scientists working in numerous disciplines. These theories have applications in many fields, and this interdisciplinary applicability may explain, at least in part, the appeal of complexity and chaos theories. Their many diverse applications give them the flavor of broadly applicable theories of basic science and basic mathematics.

***

"Is reductionism enough?" asks physicist Manoj Samal. Is it enough, he asks, for physics to take the reductionist approach, in which science seeks the truth by looking with finer and finer resolution at the basic constituents of the universe?

While Samal concedes reductionism's contributions to the success of modern science, he also puts forth that now is the time to recognize complexity as an additional frontier of science. There are three ultimate frontiers of science, Samal says—the very small, the very large, and the very complex.[212]

Samal's thinking leads him to one of Part Four's radical theories of mind physics: thought space. But more new physics first.

22.
Neutrinos

Neutrinos are an elementary particle predicted in the 1930s by theoretical physicists and observed in the 1950s by experimental physicists.

Neutrinos had to exist because of symmetry in the pattern of fermions (matter particles). If they didn't exist, we'd have to invent them.

So neutrinos seem to be a pretty well-established aspect of physics. Old-fashioned, you could say. The neutrino seems like a particle of old physics, not new physics.

Except that for decades it hasn't been clear whether or not neutrinos have any mass. And except that neutrinos have no electromagnetic charge but may nevertheless be affected by the electromagnetic force. And except that neutrinos keep popping up in the investigation of the answers to many questions of new physics.

How can mysteries of new physics still surround this long-established particle of old physics?

***

Particle physics is a great success story of physics.

An atom—ordinary matter—consists of combinations of three elementary particles: the electron, the up quark, and the down quark. So electrons themselves are elementary particles, whereas protons and neutrons are made up of elementary particles: a proton is two up quarks plus one down quark, and a neutron is two down quarks plus one up quark.

Matter is made of molecules, and molecules are made of atoms, and atoms are made of electrons and quarks. But we are not expecting that the elementary particles will in turn be found to be divisible into even more elementary particles. We think that electrons and quarks are truly elementary, although in work such as string theory we are looking to understand a more simplifying single driving concept that will further unify all particles and forces. Strings would make our understanding of the elementary particles more elegant, but there would still be a central role for the elementary particles in our understanding of particle physics.

In any case string theory jumps ahead of the story a bit, because additional elementary particles—not needed to construct molecules or atoms—have been theorized and discovered since the early days of particle physics. The muon was experimentally confirmed in the 1930s, and the neutrino in the 1950s. It is now understood, within the standard model of particle physics, that there are four types of matter particles—the electron, the neutrino, the up quark, and the down quark. And that each of these particles has a second family member or second generation or second flavor that is a heavier particle—the muon is the electron's second family member—as well as a third family member that is heavier still.

The neutrino was originally hypothesized to account for missing energy in experiments on radioactivity. The neutrino was hard to detect because it interacts so weakly. A neutrino can pass through lead a trillion miles thick without interacting with it. Neutrinos have now been detected in both particle accelerators and in cosmic rays from outer space. Frederick Reines, in 1995, and Raymond Davis and Masatoshi Koshiba, in 2002, have been awarded Nobel Prizes for physics for their successes in the detection of neutrinos.

In spite of advances in observing and understanding the elementary particles, the neutrino has been quite the troublesome problem child, with decades of experimental efforts taking place to determine whether the neutrino has mass or is massless. The neutrino has been a candidate to explain at least some of the universe's dark matter if we can determine that it has mass. Finally in the late 1990s consensus was reached among experimental physicists that the evidence supports a conclusion that neutrinos are not massless, in contradiction to what particle physics' standard model assumes. This shakes up physics and requires adjustment—perhaps a tweak, perhaps earth-shattering—to established physics.[213]

Neutrinos and New Physics

Some physicists note that the "investigation of the unknown absolute neutrino mass scale is situated at a crossing point" of new physics. They note the role of neutrinos in explaining a series of new physics phenomena: the origin of mass, the unification of the forces, the nature of dark matter, the existence of hidden extra dimensions, mysteries of the highest-energy cosmic rays and of the excess of matter over antimatter.[214] Neutrino physics is now understood to be intimately involved in both the nuclear physics of stars and the long-term chemical evolution of galaxies.[215] Neutrino telescopes probe galaxies and extra-galactic objects,[216] and roles for the neutrino are posited for the large-scale structure of the universe as well as for primordial nucleosynthesis, the universe's earliest particle formation.[217] Physicists discuss roles for neutrinos in the acceleration of the expansion of the universe,[218] as signatures of the transformation of a neutron star into a black hole,[219] and in dark energy and quantum gravity.[220] Some wonder whether neutrinos traveling faster than the speed of light—tachyonic neutrinos—create the illusion of neutrinos' mass.[221]

Neutrinos are proposed both as a tool to investigate extra dimensions and as a phenomenon of extra dimensions. Neutrino physics is used to probe both large and small extra dimensions.[222] And models in which neutrinos in our four dimensions are coupled to bulk neutrinos, or to other matter particles that propagate into the bulk of extra dimensions, are reconciled with standard-model neutrinos and their variations.[223]

What's Going On With . . . The Neutrino

The solar neutrino problem reflects discrepancies between the observed and the theoretically predicted quantity of neutrinos that are emitted during beta decay (an effect of the weak force) in the core of the sun. A related problem is the atmospheric neutrino problem, which reflects discrepancies in the quantities of neutrinos produced by cosmic ray interactions in the earth's atmosphere. These discrepancies have led to solar neutrino physics as a new, "exciting and difficult field of research for physicists." [224]

Both the solar neutrino problem and the atmospheric neutrino problem have as their solutions the discovery that neutrinos have masses and oscillate between different flavors. Like neutrinos having mass, neutrino oscillations also bring our current understanding of neutrinos beyond the standard model. The oscillatations in question are from one neutrino flavor or family to another.[225] Physicists comment: "The reality of neutrino oscillations has not really sunk in yet. The phenomenon presents us with purely quantum mechanical effects over macroscopic time and distance scales . . . In neutrinos we are confronted by real particles which behave in a quantum mechanical fashion over very large distances, 1000s of km. We are not used to this, and the physics community has yet to assimilate fully the implications."[226]

In the standard model, neutrinos interact only through the weak force. The mass and oscillation contradictions to the standard-model's assumptions regarding neutrinos have led physicists to the possibility of yet another neutrino-related contradiction to the standard model, namely that neutrinos are in fact subject to the electromagnetic force, at least indirectly, even though neutrinos have no electromagnetic charge.

Physicists propose various mechanisms through which the neutrino may be influenced by an electromagnetic field, even though the neutrino has no electromagnetic charge. These mechanisms include virtual charged particles that emerge from neutrino self-energy, oscillation phenomena, the neutrino-neutrino-photon vertex, neutrino-photon and neutrino-electron scattering, and neutrino-proton processes. These mechanisms, if they exist, would have important implications for astrophysics and phenomena of the stars.[227]

Some wonder whether the small value of the neutrino mass predicts our universe rather than the neutrino mass being explained by the universe. This is called an anthropic selection effect: the anthropic principle is frequently invoked in physics when we ask whether the conclusions we draw are dependent on who we are, not on what the physical phenomena are. Neutrinos streaming out from the universe's energy wells slow the growth of density fluctuations that are required for galaxy creation. The higher the mass of neutrinos, the fewer galaxies that will be formed. So could it be that ours is a biased view of the low mass of neutrinos, because we're looking only at a universe in which galaxy formation has not been suppressed by high neutrino mass? Could neutrino mass be a reflection of the extent to which we live in a universe populated with galaxies, a universe whose very nature has emerged due to the neutrino mass being as low as it is?[228]

***

Isaac Asimov published in 1966 *The Neutrino: Ghost Particle of the Atom,* in which he discusses his fascination with neutrinos as part of a "centuries-long chain" tracing back to the "dawn of modern experimental science."[229] And still today, articles on the latest experimental findings about neutrinos are published co-authored by over one hundred researchers working in collaboration from dozens of universities and laboratories throughout the world.[230]

And, as you might suspect, we will find neutrinos again in Part Four, as components of theorists' formulations of theories of new physics and the the mind.

23.
The Unreasonable Effectiveness of Mathematics

Scientists—biologists, chemists, physicists—have an urge to model phenomena mathematically. And mathematics is good for modeling science. Unreasonably good, you might say.

Why is mathematics so effective as a tool for science?

***

For thousands of years, mathematics has been a tool of science. The earliest astronomers modeled the cycles of the heavens mathematically, and early scientists modeled the physical world with tools of arithmetic and geometry. We saw in Chapter 2 that Roger Penrose's handful of superb theories of the physical world include Euclid's geometry from over two thousand years ago and Sir Isaac Newton's seventeenth-century models of mechanics.

"The Unreasonable Effectiveness of Mathematics in the Natural Sciences" is the title of a 1960 essay by physicist Eugene Wigner, based on a lecture on mathematical sciences that he gave the previous year. Wigner later shared the 1963 Nobel Prize for physics for his work on the atomic nucleus and elementary particles. He is also recognized in the expression Wigner's friend, based on Wigner's contemplations of the meaning of quantum physics: Wigner's hypothetical friend adds a layer of observation to Schrödinger's hypothetical cat—the cat who is quantumly both dead and alive, in quantum superposition, unobserved quantum limbo, inside an unopened box. Wigner's adding a friend to the picture allowed him to further explore the meaning of quantum reality by raising questions about when, if ever, the hypothetical becomes a universal reality, crossing interpersonal boundaries of shared observation.

In his 1960 essay, Wigner first explains that the degree of effectiveness of mathematics in the sciences is, it can be argued, unreasonable: "the enormous usefulness of mathematics in the natural sciences is something bordering on the mysterious and . . . there is no rational explanation for it."[231]

Wigner makes no claim that the observation of mathematics' effectiveness is an original thought of his own. He notes the statement attributed to Galileo that "the laws of nature are written in the language of mathematics."[232] And he notes that Einstein, like many physicists and other scientists, saw aesthetics in mathematics and tested theories by these aesthetics: "the only physical theories which we are willing to accept are the beautiful ones," remarked Einstein.[233]

Wigner asserts, by example after example from the history of physics, that the language of mathematics is the correct language for physics. He proposes as an empirical law of epistemology—the nature of knowledge—that the physical world is of a mathematical nature: "the mathematical formulation of the physicist's often crude experience leads in an uncanny number of cases to an amazingly accurate description of a large class of phenomena."[234]

Wigner also discusses the possibilities for mathematical formulations of phenomena at the heart of this book—the reconciliation of quantum physics with general relativity, and the phenomenon of consciousness.

Ultimately, Wigner prefers to observe, to accept, to appreciate, but not necessarily to explain the effectiveness of mathematics in science: "The miracle of the appropriateness of the language of mathematics for the formulation of the laws of physics is a wonderful gift which we neither understand nor deserve. We should be grateful for it and hope that it will remain valid in future research and that it will extend . . . to wide branches of learning."[235]

In Part Four, the credibility of our Number One Radical Theory of New Physics is, for some, greatly enhanced by its mathematical aesthetics: it deploys mathematics' only two complete mathematical systems, using one system to model the physical world and the second system to model the world of the mind, uniting the two through a correspondence that connects quantum physics and the mind. This mathematics is unreasonably effective at modeling new physics and the mind.

A 2002 Application

The links between science and mathematics have continued unabated. In 2002, physicist Stephen Wolfram published *A New Kind of*

Science, in which he emphasizes the far-ranging implications of even the simplest mathematical rules. Wolfram reports on his exhaustive study of some simple mathematical rules and shows that, from these rules, very simple computer programs can generate very complex results.

A New Kind of Science includes extraordinary displays of tile-like graphics and of graphs that derive from simple rules about patterns of numbers and about the colors of cells or cells' neighbors. Wolfram points out that usually simple rules result in simple, repetitive behavior, but occasionally they yield nested patterns and more complex results. Wolfram sees these more complex patterns in the world's arts and ornamentations. He classifies his findings and draws conclusions about the foundations of mathematics, cryptography, statistical analysis, computer science, technology, even financial systems.

But Wolfram also takes this much further, to the growth of crystals, fluid flow, the growth and pigmentation of plants and animals, visual and auditory perception and analysis, evolution of life forms, the relationship of space and time, causality and sequencing of events, reversibility and irreversibility, relativity, elementary particles, gravity, quantum phenomena, and models of the universe.

In fact, Wolfram's complexity theory—his new kind of science—is based on the principle that everything in the universe derives from rules that can be represented by very simple computer programs.

Wolfram's urge is familiar. Mathematics has an aesthetic quality that underlies its organizing principles and calls out for application to the universe's large and small.

We'll see more mathematics in Part Four's speculations on new physics and the mind. We'll see theories, like Wolfram's, that draw conclusions of great breadth from mathematics that is simple. These theories gain some of their intuitive appeal from their being reducible to simple mathematics. And like Wolfram's theories, these are not mainstream. We await the passage of time and effort to conclude which stream is flowing in the right direction.

***

A T-shirt popular at M. I. T. has written on it Maxwell's four equations of electromagnetism:

$$\Delta \cdot \mathbf{E} = \rho$$
$$\Delta \times \mathbf{B} - \partial \mathbf{E}/\partial t = \mathbf{j}$$
$$\Delta \cdot \mathbf{B} = 0$$
$$\Delta \times \mathbf{E} + \partial \mathbf{B}/\partial t = 0$$

In a typical variation of the T-shirt, these mathematical statements of Maxwell's equations—equations of light and other electromagnetic phenomena—are supplemented with the words: "And God said: Let there be light!"

Even M. I. T. students can be fascinated by the power and mystery of mathematics. And maybe also a bit proud that (at least for the midterm exam) they understood these equations.

But perhaps more fascinating is a folluw-up fact that students learn. All four of these equations can be compressed into just one very brief formula:

$$\partial \mu F^{\mu\nu} = j^{\nu}$$

Typically, a physicist or mathematician will gain a great deal of comfort about the validity of any complex theory—such as Maxwell's theory of the electromagnetic force—by virtue of its being reducible to a few mathematicial symbols on a single line.[236] This is a measure of the power we feel mathematics holds over the physical world: mathematical simplicity, mathematical elegance, give us comfort that we've uncovered a deep truth and understanding.

24.
The Myths of Time and Mass

Scientists and mathematicians often exhibit an urge to simplify. It's not enough to explain something; the explanation needs elegance to be fully satisfying.

Bringing elegance into a discussion of science may sound odd, but scientists and mathematicians do juxtapose a proof or demonstration that is accomplished merely by brute force with one that is truly elegant. An elegant proof proceeds simply, straightforwardly, with as few steps as possible. It is not an exhaustive listing of the facts and possibilities, showing one by one that each possible route leads to the premise being proven. It is a work of artistry and economy, with no needless embellishment.

This leads some physicists to ask: Can physics be simplified to remove the need for a concept of time? A concept of mass?

On the surface, of course, this is ridiculous, given how central are the roles that these two properties play both in our everyday lives and in the methods of science over the millennia. But both concepts have bothered some scientists, although in different ways.

Let's hear the arguments for a more elegant understanding of the universe, an understanding that relegates time or mass to the ashcan of history, archaic and unnecessary illusions, embellishments that only get in the way of understanding what's really going on.

***

The End of Time

"Physicists are using too many concepts," says theoretical physicist Julian Barbour in his 1999 book, *The End of Time: The Next Revolution in*

Physics.[237] And for Barbour, time in particular represents one concept too many.

Barbour goes right to the heart of the matter by proposing a vision of the physical world in which time is a secondary, derived concept, not a basic, real phenomenom. Yes, you do think you experience the flow of time. But consider this more deeply and you can envision a structure of the world in which time does not need to exist.

Barbour's world without time is instead a configuration space of nows. The nows that we experience today, the nows we experienced yesterday, the nows we will experience tomorrow—all these nows exist forever in a timeless realm which Barbour calls Platonia. Barbour names this realm based on Plato's philosophical approach in which ideas are real, forms are real, but our constrained physical existence permits us only the most limited views of only the shadows of these ideas and forms. In Barbour's construction, our consciousness, experiencing different nows, gives us an illusion of the moving present.

And for Barbour it is not just time that is an illusion, a derived concept. Space, too, is an artificial construction. "Only things exist," says Barbour; "the supposed framework of space and time is a derived concept, a construction from the things."[238]

Barbour uses his framework to help resolve modern physics' great mystery—how quantum physics is to be reconciled with general relativity. Work of Austrian physicist Ernst Mach, famous today for the Mach numbers that represent multiples of the speed of sound, is used as a centralizing concept. Mach's principle, developed during the second half of the nineteenth century, states that an object's mass is created only in relation to all of the mass in the universe. Barbour draws conclusions from this principle that make things primary and space and time derived. "The demise of space and time is inevitable."[239]

Barbour's *End of Time* echoes other theories and world views. In the strong determinism view, for example, all of the universe's history—from the big bang onward, including our past, our present, and our future—is already in place and determined. And strong determinism has its variation, the many-worlds interpretation of quantum physics, in which there is a world for all paths of quantum resolutions.[240] Certainly we will see more echoes later in this book, as we discuss in detail geometrodynamics, which Barbour refers to as close to his own interpretations.[241]

What these views have in common is a diminishment of time's role in physics, in favor of a view of the universe including at this moment all that is past, present, and future.

What's Going On With . . . Other Radical Views of Time

There are several variations of models of the physical world in which time has two or more dimensions.

When a time travel plot device is used in science fiction, the paradox that needs to be dealt with is typically a paradox of causality—killing your grandparent, for example, before a parent is born. In one theory, timelike extra dimensions prevent these causality violations by permitting the violating actions only in alternate time dimensions. These alternate dimensions may not yet have been noticed by us because their magnitude is smaller than the Planck time or length. Or perhaps there is some mechanism through which particles are localized in the extra time coordinate, without the possibility of communication with our standard time dimension.[242]

In another proposed physical model, hypertime extends string theory beyond M Theory's eleven dimensions. M Theory unifies the string theories, and this hypertime theory takes this further—to an F Theory with twelve dimensions—by hypothesizing a second time dimension that helps resolve mysteries unanswered by M Theory.[243]

The Problem with Mass

In the jargon of physicists, the problem with mass is: why is inertial mass the same as gravitational mass? Earlier, we encountered this problem briefly as part of the explanation for why it is so difficult to find a theory of quantum gravity.

Inertial mass is the quality of an object that relates force to acceleration. When an object's inertial mass is large, it takes a lot of force to accelerate it, to make the object move faster, or even to set it in motion at all (move it from zero velocity to a higher velocity). But get out of the way once you've got a heavy object in motion: an accelerating heavy object has a lot more force behind it than a lighter object accelerating at the same rate.

We have Sir Isaac Newton's second law of motion: $F = m\,a$, force equals mass times acceleration. Here, mass can be filled in after the fact, a constant that relates force to acceleration. An object's mass—inertial mass, that is—is determined by applying force to the object and noting how much acceleration results. Divide force by acceleration, and you've determined the object's mass.

Separately, we have a completely different understanding of mass: mass tells us how heavy an object is, how much "stuff" it contains. And how large an effect (gravitational pull) it will have pulling on another object. Not how resistant it is to being put into motion. Not how much force is behind it once it is in motion. But how heavy it is.[244]

At this point, your reaction to physicists' concerns may well be: get a life! It just doesn't occur to us in our everyday lives that inertial mass and gravitational mass are two distinct concepts at all. We take mass for granted, in both its inertial and gravitational senses.

But many physicists have a better idea.

The better idea is the Higgs boson. This is a theorized (never observed) elementary particle that has the function of giving matter its mass.

The quest is on to observe the Higgs boson through use of particle accelerators. The theorized size of the Higgs boson is too small for it to be observed in today's accelerators, and this quest is as yet unfulfilled. But the theory of the Higgs boson is extensive and very well established—it explains a lot, and provides an answer to the bothersome questions surrounding what gives matter its mass.

In fact, although never observed, the Higgs boson has wide acceptance among physicists. It is part of the standard model of particle physics, theorized as producing a field that fills the universe and creates mass in any particle that has mass. In the standard model, the Higgs boson is the origin of all mass.

A minority of physicists reject the existence of the Higgs boson, driven largely by its great inelegance. For example, Leon Lederman, who shared the 1988 Nobel Prize for physics for work on particle physics, has written with associate Dick Teresi the book *The God Particle: If the Universe Is the Answer, What Is the Question?* The God particle of the title is the Higgs boson. Regarding the standard model, Lederman and Teresi comment: "It's not simple enough . . . There are too many parameters in the standard model, too many knobs to twiddle."[245] Lederman and Teresi cite the introduction of the Higgs boson into the standard model as one of the standard model's flaws, and they quote Martin Veltman— one of the developers of the theory of the Higgs boson—as calling it "a rug under which we sweep our ignorance."[246] Lederman and Teresi also note the even more pointed comment of Sheldon Glashow, who shared the 1979 Nobel Prize for physics for developing the concepts of the standard model's electroweak force: Glashow refers to the concept of the Higgs boson as "a toilet in which we flush away the inconsistencies of our present theories."[247]

To provide a flavor for why the Higgs boson is viewed as so inelegant by this physicist minority, we need to digress briefly into the world of particle physics to discuss the other bosons.

In the standard model of particle physics, there are two types of particles—matter particles (fermions) and force-carrying particles (bosons).

There is enormous symmetry in the organization of characteristics of the standard model's matter particles, so much so that great efforts have been made to confirm the existence of certain matter particles (for example, the hardest-to-detect quarks) simply because these symmetries require that the undiscovered particles have to exist. These matter particles (fermions) share the characteristic that they all have "spin" of ½, spin being a quantum property similar to rotational symmetry.

The bosons that carry physics' four forces are all characterized as "gauge bosons" ("gauge" meaning measure). The standard model's gauge bosons don't have the extent of symmetry of organization that the fermions have. And the troublesome graviton—which transmits the force of gravity—stands out by having a spin of 2, unlike the other gauge bosons (photons, gluons, W and Z bosons), which transmit the electromagnetic, strong, and weak forces and have spin of 1.

That the graviton stands out as exceptional does not surprise us— gravity is at the heart of modern physics' central mysteries. But the Higgs boson is in yet another category of its own. It is not a gauge boson, meaning its force does not conform to the mathematical structure of the other bosons. It has 0 spin. And its field (the Higgs field) is undetectable and differs from the other bosons' fields in that it has no direction.

Marble, Not Wood

Albert Einstein spent much of the second half of his life trying to create marble from wood. For Einstein, general relativity was marble— the beauty and elegance of the geometry of spacetime, whose curvature creates the effects of gravity. But his theory of matter—even his $E = mc^2$ that connects energy and mass—was wood, inelegant and not geometric in origin.

Einstein never made marble from wood. He never created a unified theory of the forces, and never geometrized matter.[248] But others have continued this quest.

What's Going On With . . . Theories of Mass

One important direction of this quest is the induced-matter interpretation of mass.[249] In this interpretation, which University of Waterloo physicist Paul Wesson and others have been developing since the mid-1980s, the features we recognize as those of matter (such as density and pressure) result from the geometry of a fifth dimension. This is truly Einstein's marble: mass becomes a separate coordinate within the geometry of spacetime-mass.

Others look at the possibility of matter being formed in the early universe as a result of five-dimensional spacetime being squeezed down into four dimensions. In this formulation, the five-dimensional universe was a vacuum, dominated by radiation, but as the fifth dimension was compactified the four-dimensional world became dominated by matter, which was formed from radiation. A variation of this theory accommodates not just five dimensions, but the ten or eleven dimensions needed for string and M theories. Another variation introduces a speed of light that was variable, not constant, in the early universe in order to explain additional cosmological observations.[250] This theory and its variations all differ from Wesson's induced-matter model, in that in this theory matter physically derives from the compactification of extra spacetime dimensions, whereas Wesson's theory is a geometrical formalism explaining how four-dimensional matter can be understood as Einstain's marble if we consider matter in five-dimensional geometry.

Other physicists propose a different extradimensional route to explain mass without the Higgs mechanism. These physicists propose a model in which force-carrying particles in the extra dimensions produce mass in our familiar four dimensions, without the need to introduce the Higgs boson. Their model also addresses the hierarchy problem of the wide range in strengths of the forces.[251]

And in yet a different approach, fermions (matter particles) are an alternative to the Higgs boson. This approach looks at the symmetries in the organization of the characteristcs of the matter particles, which the standard model extends to three generations or families of symmetry. This approach explores the possibility of a fourth (even heavier) generation of elementary matter particles, and suggests that such an additional set of fermions could offer an alternative to the Higgs boson.[252]

In the closing chapters of this book, as we look at some of new physics' more radical theories, we'll encounter topological geometrodynamics (TGD), a geometry-based theory of everything. Creating a mathematics-based model of the Higgs mechanism was actually a very early effort on the part of TGD's developer, Finnish physicist Matti Pitkänen. But this modeling eventually led Pitkänen far afield, toward radical new applications. And in these applications, about which we'll be hearing much more, Pitkänen disavows "Higgs trickery."[253]

Perhaps farthest afield are possibilities that can be consided in the context of studies of neutrino oscillations. As we discussed in Chapter 22, investigation of neutrino oscillations has been a primary methodology in confirming that the elementary particle neutrino is not massless but does in fact have a small mass. These investigations do not actually permit us to prefer a conclusion that neutrinos have mass over a conclusion that neutrinos are tachyons—that they travel faster than the speed of light. Even though tachyons remain just hypothetical, this hasn't prevented physicists from considering how they would behave, if they exist. And one of the properties of tachyons is that they travel backward in time. So perhaps the difficulties of distinguishing the properties of mass from the properties of tachyons may lead to development of a deeper relationship between time and mass.[254]

***

The "myth of mass," for some physicists, is a search for a more elegant understanding of matter and of mass than the understanding that physics' standard model now provides.

And the "myth of time," for some physicists, goes even further, toward an "end of time," a removal of time—and even space—from the basic concepts that are needed to create a theory of everything.

That time and mass may be just "myths" will not be taken whole into the closing chapters of this book, in which we look at some physicists' radical new ways of approaching physics. But the drives to question the accepted understandings of time and mass are part of a trend, a growing list of what's troublesome about today's canonical approaches to physics.

25.
The Role of Art

Art gives us deeper meaning, gives dimension to life, some say. (Like new physics' extra dimensions?)

Quantum physics' being at the same time as only having the potential to be. (This has never been a problem in the world of art.)

A photograph capturing a moment in time. (Resolving physics' questions about the nature of time and its arrow?)

Social commenters connecting travelers' fascination with air travel to their hope for "perpetual self-renewal through forward motion."[255] (Special relativity's slowing of time and aging, affecting the mind through the biophysics of the brain?)

This is all just cute wordplay, right?

Well, a handful of scientists say no. A handful of scientists look for connections between art and physics.

Art is, after all, created in our physical world. And if nothing else, art is all about provoking novel reactions, new ways of sensing the world, of realizing our psyches, transforming us. And maybe artists, those of the right brain, do have something of value for the world of left-brained science.

***

Metaphorms

Perhaps the most elegant and complete advocate for art's role in science is Todd Siler, author of *Breaking the Mind Barrier: The Artscience of Neurocosmology*. The cover of this 1990 book is captioned "a brilliantly original way to think about art, science, the mind, and the universe," spelling out in four plain-English words the two neologisms of the book's subtitle, artscience and neurocosmology.

"It is through art that we see these two sides of the same interpenetrating matter—brain and cosmos—facing each other," says Siler.[256] So it's the mind and the universe, the brain and the cosmos, that Siler is connecting. Art and *metaphorms* are the central methods of inquiry.

Metaphorms—art-based metaphors between brain and universe, between art and science—appear throughout *Breaking the Mind Barrier.* Throughout history, Siler would argue.

A sacred mandala pattern of ancient Buddhism metaphorms to a potassium-intensity contour map of Betelgeuse, a star in the constellation Orion, 650 light years from earth.

Two paintings overlap as Two Planes of Thought Intersecting, collages of what is seen and what remains to be seen in the hidden territory of the mind. Artwork of the state of mind. Artwork of thought assemblies. Art as A.R.T.—All Representations of Thought.

Siler imagines a *cerebrarium*, a multimedia artwork that responds to the imaginings of its human visitors, a learning arena that through metaphorms learns as the visiting humans learn.

Brain and star are processmorphs—alike in process, although unlike in form or appearance. The brain and cosmos have evolved in parallel. Local superclusters of the brain metaphorm to local superclusters of the heavens. The neural tissue of the brain's hippocampus metaphorms to spiral galaxies. Our nervous system's neural winds metaphorm to the earth's magnetosphere. Lightning bolts are nature's neurons.

Look at the inventor and what he or she invents. "Do the principles of fusion-fission reactors and particle accelerators reflect the working of the human nervous system?" Siler asks.[257] This leads to the concept of *cerebreactors*, to the cerebral fusion of intuition and the cerebral fission of reason. Siler lists over one hundred human inventions as processmorphs of the human brain.

Siler's neurocosmologists decode the brain and decode the universe. The universe, with its hundreds of billions of galaxies, each with its hundreds of billions of stars. And the brain, with—as Siler notes—seven billion neurons coordinated together with nine thousand miles of fiber per cubic inch.[258] Siler is clear: galaxies communicate by means of gravity, humans communicate by means of consciousness. Consciousness metaphorms with the force of gravity.[259]

So where does Siler take his metaphorming of neural and stellar systems, his brain/universe connections? Transcending shifts in scale, metaphorically pointing out similarities in the structure and dynamics of brains and stars, Siler takes us to expanding visions of our natural intelligence, to networks of creation, to recalculating the importance of the arts. He maps the general system of the neurocosmos.

Siler's view is far beyond similes and metaphors between art and science. It is not simply that we see pictures of science in art and pictures of art in science. Instead, it is an affirmative embrace of art—the human creation—as the link between the brain and the universe, between the mind and the cosmos.

More Art in Physics

Ezra Pound, twentieth-century American poet, proclaimed that artists are the antennas of the human race. So it may be inevitable that we find art as we look for the secrets of the physical world.

In *Einstein, Picasso: Space, Time, and the Beauty that Causes Havoc*, Arthur I. Miller finds parallels in the personal and working lives of Albert Einstein and Pablo Picasso, and extends this to a parallelism between Einstein's relativity and Picasso's cubism. Born just two years apart (Einstein in 1879, Picasso in 1881), Einstein and Picasso were central to the trends toward abstraction and new forms of imagery common to both science and art during the twentieth century. Einstein and Picasso mark the end of the classical world, Miller notes, and the beginning of new physics and new art.

Parallels between science and arts are also elaborated in Douglas Hofstader's Pulitzer-prize-winning *Gödel, Escher, Bach: An Eternal Golden Braid*; in research psychologist Lawrence Le Shan and physicist Henry Margenau's *Einstein's Space and Van Gogh's Sky*; in Leonard Schlain's *Art in Physics*; in *Inside the Mind of God: Images and Words of Inner Space*, edited by Michael Reagan.

Art plays an implicit but important role in gauging the success of theories of physics. We gain comfort when a theory demonstrates supersymmetry, and we are especially comforted by theories' elegance and simplicity. Alan Lightman, in his introduction to Alan Guth's 1997 book *The Inflationary Universe: The Quest for a New Theory of Cosmic Origins*, says simply: "Unification and simplicity have been the eternal Holy Grail of physicists and artists."[260]

This is, at least in part, a manifestation of a broader theme that it is two universes that we inhabit—the universe of the cosmos and physical world, and the universe of the mind. Timothy Ferris, in *The Mind's Sky: Human Intelligence in a Cosmic Context*, fully embraces the parallels between these two universes, all the way to invoking a human drive—our destiny—to communicate with the universe's other intelligences. Ferris sees neuroscience intersecting with the search for extraterrestrial intelligence, and he finds a central nervous system—a galactic thinking network—spread throughout the Milky Way.

Peter Pesic, in his 2002 *Seeing Double: Shared Identities in Physics, Philosophy, and Literature*, sees quantum physics as the key to answering how individual identity emerges from perfectly nondistinct elementary particles, particles with no individual identity at all.

The proponents of *ethnomathematics* find cultural imprints even in mathematics.[261]

Leon Lederman and Dick Teresi note, in their 1993 book *The God Particle: If the Universe Is the Answer, What Is the Question*, how frequently general-audience science books end with philosophical musings on our role in the universe. With tongue only partly in cheek, they close their book with their own "obligatory God ending."

And Then There's Physics in Art

Quantum physics has been making an appearance in art.

We have Michael Frayn's popular play "Copenhagen," about Werner Heisenberg's 1941 visit with Niels Bohr and Bohr's wife. Here quantum physics is both subject and metaphor. What is the accurate version of what was said when these three met? There are parallel paths of possibilities. Were Heisenberg's activities—in support of Nazi Germany's efforts to develop the atomic bomb—everything they seem? Correspondence is drafted, rewritten, unsent and sent. Uncertainty abounds, and there are limits on what can be known.

Architectural critics review proposed buildings in the context of "the culture of time and space". A building looks like a spaceship, but not just a metaphorical spaceship. "Buildings don't move in space. Like civilizations, however, they move in time. Seen in this dimension, . . . [the] graphically dynamic contours . . . [are] more than metaphoric. They . . . [are] spaceships expressly made for moving along the arrow of time."[262]

Thus the architecture resonates for us because it reaches to reconcile time and space, to resolve core concepts of physics. A deeply felt quest for scientific knowledge resonates with our sense of artistic appreciation.

Charles Ross has spent twenty-six years building "Star Axis," an art installation in New Mexico. When completed in 2006, this installation will extend a quarter mile across and eleven stories high. Visitors walking through it will experience the earth's movement through space and time due to the installation's alignment with the earth's rotational axis. This "cocktail of science and art" brings to earth and to human scale the geometry of the heavens, and gives the visitor "a spiritual connection with the universe."[263]

Lynn Gamwell, in *Exploring the Invisible: Art, Science, and the Spiritual*, finds that modern science has profoundly transformed modern art. Gamwell finds inspiration for artists in scientific images from the mid-nineteenth century on. *Exploring the Invisible* is filled with magnificent images of microorganisms, fossils, undersea creatures, spectrums of light and sound, atomic particles, images from telescopes and microscopes, that, Gamwell finds, have inspired artists' subjects, techniques, and modes of representation. As Newton's clockwork universe evolves to Einstein's spacetime, Gamwell finds the evolution to abstract art, to impressionism, surrealism, postmodernism.

Fiction

The world of fiction is a special form of art, bringing language and writing of our analytic left brains into the sphere of creativity.

Modern fiction is surrealism and neo-surrealism. It is magical realism, in which the fantastic is ordinary. It is metafiction, such as Margaret Wander Bonanno's *Preternatural* series, in which the protagonist discovers "that the science fiction novel about telepathic aliens she thought she was writing was actually being dictated to her by telepathic aliens who thought she was a fictional character in one of their stories."[264]

Science fiction has given us views—sometimes science-based, sometimes not—of black holes, of extra dimensions, of parallel universes, of teleportation, of time travel.

But science fiction is a product of we of the physical world. When science fiction resonates with its readers, could this be because it's striking a deeply real chord?

Quantum physics is emerging as a popular dramatic convention in fiction. And the vocabulary of physics has seeped into how fiction is discussed, reviewed, read, written. A family "grapples with . . . relativity"—with "curves in time" and "violation in causality"—as history resonates in the present.[265] A book is "organized topographically,"[266] and another—Margaret Visser's *The Geometry of Love: Space, Time, Mystery, and Meaning*—reflects "the interaction of architecture and prayer through time."[267] Another novel employs an unexplained dramatic convention of "tumbling back" to transport its characters through time.[268] And another uses the conventions of quantum physics to explore connectedness of widely separated simultaneous events.[269]

Susan Strehle, in *Fiction in the Quantum Universe*, puts forth that the postmodern abandonment of realism is not an abandonment of reality. This is because quantum physics' reality has taken us beyond realism to *actualism*—active, dynamic, actual. The narrative involves the observer.

Positions in time and space are uncertain. "Narratives can't steer by the fixed poles that guided realistic fiction."[270] Strehle discusses quantum concepts' parallels in the works of six contemporary authors—their "actualistic inconclusiveness,"[271] their "mixed choices and plural aims."[272] Strehle finds in their work fiction that is discontinuous, statistical, energetic, relative, subjective, uncertain. Fiction that deregulates time. Fiction that energizes space.

In Italian essayist Italo Calvino's 1968 fabulist fantasy *Cosmicomics*, the characters' beginnings date to the big bang. These characters are actually mathematical formulas and particles of elementary physics, chemistry, and biology. They exist within nebulae as the universe expands and as the sun and planets form. When they were kids, their only playthings were hydrogen atoms. They note where they are with respect to the rest of the Milky Way so that they can compare the scene in two hundred million years when they'll get this view again. Although there was only one cleaning woman for the whole universe at the point of the big bang, she had nothing to do because there was no room for dust.

Even physicists get in on the act. *Einstein's Dreams*, by Alan Lightman, physics and writing professor at M. I. T., incorporates multiple senses of time, riffs on simultaneity and ordering. The journal *Physics Today* publishes physicist Gordon Kane's fictional story about particle accelerator confirmaton of extra dimensions.[273] And physicist Sheldon Glashow postulates the "UBS" network of the future—Universal Broadcasting System, with communications from throughout the Milky Way and beyond. Consult your local listings for this special: "For the serious listener, there will be a panel discussion with more than 50 participating galaxies on the subject of the mysterious dark mass of the universe."[274]

It is not just the intersection of fiction and physics that is under scrutiny, but also the intersection of fiction and consciousness. David Lodge discusses ways that novels represent consciousness and how these representations have changed over time. He notes the transition from early fiction's style of the dominant voice to free indirect style, in which the authors speak the minds of their characters. Lodge traces how the balance changes back and forth in classic, modern, and postmodern works between dialogue and narrative—between direct speech and the rendering of characters' unspoken thoughts—finding parallels with evolving understandings in science and philosophy. And fiction continues to evolve, Lodge says, as technology changes us all—film, cyberspace, virtual reality.[275]

Why Stop Here?

Religion comes from us, we of the physical world. And so do all of our paranormal observations and fantasies.

Can we divorce all this from the physical world?

Can we be certain that religious beliefs—deeply, sometimes fervently, held—have no connection to physical reality?

And is every paranormal phenomenon an irrational fantasy, with no basis in physical reality? Extrasensory perception? Sightings of spectres and extraterrestrials? Hidden voices?

And what about dreams?

What about the chemistry that is noticeable among a well-functioning group of people?

Did Rudolph Valentino "forever change the electric charges of both men and women" through his starring role in the movie *The Shiek*?[276]

A story's narrator "spends evenings wandering the grounds of a temple, her mind floating through a world gone pleasantly abstract."[277] This is an essentially precise scientific description of the mind as it is formulated in a theory of physics we'll discuss in Part Four.

In *The Dancing Wu Li Masters: An Overview of the New Physics*, Gary Zukav notes that, in Taiwan, physics is called wu li. Wu means either matter or energy. Li means universal order or universal law or organic patterns. So wu li—physics—is all about patterns of organic energy. Zukav relates this to ancient Zen Buddhist and Hindu concepts.

Why do we have a sense that ancient or "primitive" people have been closer to the spirits of the universe than civilized moderns are?

This path has been traveled by others. Fritjof Capra published *The Tao of Physics* in 1975. And perhaps you've seen a full-page newspaper ad in which Maharishi Mahesh Yogi promotes his proposal for permanent world peace. This proposal searches different levels of physics in the quest for the invincibility that is the key to world peace. Maharishi presents his world view: This invincibility is not to be found at the level of specific applications of physics, such as electronics, telecommunications, or computer science. Nor in the theories—classical mechanics, thermodynamics, acoustics, optics—which support these applications. Nor in groupings of these theories—nuclear physics, quantum mechanics, atomic physics. Nor even in physics' forces, its matter particles, its grand unification, gravitation, lepto-quarks. No, Maharishi tells us. Invincibility is found only at super-unification, the unified field of all the laws of nature, the theory of everything. This is pure transcendental consciousness, fully awake within itself, transcendental to space and time. This is Maharishi's invincibility—invincible everywhere and for all time. It's also a summary of the history of the progress of twentieth- and early-twenty-first-century physics.[278]

In their new translation of Genesis, Rabbis Meir Zlotowitz and Nosson Scherman translate the first few words of Genesis as: "In the beginning of God's creating the heavens and the earth—when the earth was astonishingly empty . . ."[279] They discuss the words astonishingly empty—in Hebrew, two words, *tohu* and *bohu*. The authors note that the words—sometimes translated desolate and void—are difficult terms, "laden with esoteric connotations."[280] And they cite the mystical interpretation of these words by the twelfth-century Jewish scholar Maimonides. Maimonides interprets *tohu* "as being the very thin substance—entirely devoid of form but having potential—which was the primary matter created from absolute nothing by God . . . It was from this . . . that He then formed and brought everything else into existence, clothing the forms, putting them into finished condition . . . The form which this substance finally took on is called" *bohu*.[281]

Does Genesis anticipate by millennia our modern understanding of the emergence of energy/matter at the big bang?

***

Can physics really end with particles and forces?

If we bring into physics art, religion, the mind, and consciousness, where will we ever stop?

Do we open this door at all? And if it's a door marked "science only," will there be anything to see?

26.
The Fine Structure Constant

The number 137 has near-Kabbalistic meaning for some physicists.

Kabbalah is Jewish mystic interpretation, which includes numerologic aspects.

Max Born, an early quantum physicist and winner of the 1954 Nobel Prize in physics, wrote a 1936 paper "The Mysterious Number 137."[282]

The number represents the fine structure constant, a fundamental parameter of physics, relating electromagnetic force strength to spacetime quanta.

Hundreds of physicists have studied and continue to study the fine structure constant and try to explain its meaning.

***

The fine structure of an atom's electromagnetic spectrum is demonstrated in the atom's characteristic spectral lines. The spacing between these spectral lines reflects small effects of energy levels associated with the spin of the atom's electrons. These spectral lines are an important tool for determining the composition of chemical compounds. For example, spectral analysis permits inference of the chemical makeup of distant astronomical bodies.

The fine structure constant, for which physicists' symbol is the Greek letter α (alpha), has been derived as a simple relationship among three important constants of nature: the speed of light, the charge of an electron, and Planck's constant, which gives us the smallest size of quantum physics' quanta. The fine structure constant is used in predicting the spacing of spectral lines, and its value is very close to the fraction 1/137. Because fractions can be awkward, physicists often prefer to talk about the reciprocal of the fine structure constant, whose symbol is α^{-1} and whose value is very close to 137.

Since early in the twentieth century, physicists have been intrigued by this number, and it is considered a fundamental and universal constant with deep significance.[283]

The fine structure constant has now been determined to many digits of accuracy (α^{-1} is just a bit larger than exactly 137), and physicists publish papers on how this exact value can be determined from other basic quantities. Some propose a basic relationship between prime number theory and the exact value of α^{-1},[284] and others have derived α^{-1} by using only the numbers 1, 2, 3, 4, and geometry's π.[285] Numerous special applications of the number 137 are noted in arithmetic, algebra, and geometry, leading some to conclude that "some sort of algebraic geometrical constraint exists in Physics which selects the number 137."[286]

Analysis of the fine structure constant has been elaborately extended to work on the fractal structure of spacetime, to the M Theory generalization of string theory, to physicist Roger Penrose's geometry of tilings and to Penrose's twistor theory of quantum gravity, and to the *p*-adic mathematics that will figure into some of Part Four's speculations.[287]

Other physicists also relate the fine structure constant to various physical and cosmological phenomena, including dark matter and dark energy,[288] inflation,[289] gravity,[290] and the energy associated with the recoil of an atom that has absorbed a single photon.[291]

Recently, evidence has developed that the fine structure constant has increased slightly since the early stages of the universe, and this has set physicists to explain the implications for cosmology and particle physics.[292]

***

The intrigue surrounding the fine structure constant captures new physics' themes of the search for unifying concepts, of mathematics' implications for the physics of the universe, and of openness to radical reconfigurations of our understandings of even the most elementary concepts of science.

Part Four
Speculations

In this, the fourth and final part of *New Physics and the Mind*, we play physicist. Quantum physicist. Relativistic quantum physicist. New physicist of the the mind.

How about a "top ten" list of theories of new physics and the mind, theories that go beyond the standard model?

Who are our role models? Who among physicists has already traveled this path?

And where in the universe does this take us?

***

Part Four Summary

How have physicists speculated about the phenomena of new physics, and about the nature of the mind? Part Four counts down ten physicists' speculations, slowly introducing strands and themes that the later theories test, reject, refine, and ultimately wind into fully developed theories of everything, including all of new physics as well as the physics of the mind.

Chapter 27. Goyaks: Beyond the Standard Model. An obscure theory of physics proposes goyaks—the Armenian word for existence, or an existing structure—as a central organizing phenomenon. This theory is presented as a paradigm for hidden radical theories of new physics that have been proposed over the years in contradiction to physics' standard models. This chapter discusses the many aspects of new physics that physicists wonder about, that physicists see if they can incorporate within the standard models or extensions of the standard models. If they can't, how radically must the standard models be reconfigured?

Chapter 28. Hidden Physics Countdown: Ten Radical Theories of New Physics. Ten largely obscure theories will be discussed which radically reconfigure standard understandings of physics. This chapter discusses the selection methodology and presents some theories not included in the Top Ten. The central, and radical, thesis of *New Physics and the Mind* is that a new direction for science is obtained from theories that combine elements of the physics of consciousness with elements of new physics. Theory #1 combines the physics of consciousness with every aspect of Part Three's new physics.

Chapter 29. #10 Olavo's Rederivation of Quantum as Classical. Olavo proposes that quantum theory is superfluous and that we can understand all of physics without quantum theory.

Chapter 30. #9 Makerník's Geometric Description of Quantum Physics. Majerník proposes a geometric model, involving extra dimensions of imaginary spacetime, that explains the quantum phenomenon of multiple possibilities that proceed, unobserved and in parallel, until collapse by observation.

Chapter 31. #8 Avery's Dimensional Correspondence. Avery proposes that our five senses correspond to various configurations of dimensions of space and time. He draws far-reaching conclusions from this framework.

Chapter 32. #7 Botta Cantcheff's Phenomenological Spacetime. Space and time are not passive backgrounds against which physical phenomena take place. Instead, space and time are actively constructed from the processes of the physical world.

Chapter 33. #6 Spaans' Topological Dynamics. Spaans models quantum superposition and the quantum jump through a topology for the universe's ground state that is linked to an excited state's lattice geometry. Through this model, Spaans explains physics' elementary particles and forces, and proposes a framework for quantum gravity.

Chapter 34. #5 Kafatos and Nadeau's Conscious Universe. Consciousness is a universal phenomenon of the physical world.

Chapter 35. #4 Sirag's Reflection Space. Reflections within extradimensional crystal structures explain physics' elementary particles and forces, quantum phenomena, and consciousness.

Chapter 36. #3 Sidharth's Quantum Black Holes. Physics' elementary particles and forces are phenomena of black hole topology. Fluctuational cosmology and quantized fractal spacetime explain important aspects of the universe.

Chapter 37. #2 Samal's Thought Space. The mind is a region of extradimensional space, acting in geometric concert with the physical world.

Chapter 38. #1 Pitkänen's Topological Geometrodynamics. Physical spacetime and classical physics are physical representations of a quantum world linked to the mind. The mind operates according to a mathematical system that is linked to, but is radically different from, real mathematics. Both the real physical world and the world of the mind are constructed as hierarchically linked four-dimensional universes—many-sheeted spacetime. Our biology links us to our minds and to the universe.

Epilogue. Type 1, 2, 3 Worlds. Cosmologists consider the possibility of intelligent extraterrestrial life, and categorize possible stages of advancement for extraterrestrial civilizations.

Afterword, by Matti Pitkänen. Physicist Matti Pitkänen, the developer of TGD, topological geometrodynamics, discusses new physics, the mind, and TGD.

27.
Goyaks: Beyond the Standard Model

Goyak is the Armenian word for existence, or an existing structure.

Physicist G. T. Ter-Kazarian, working at the Byurakan Astrophysical Observatory in Armenia and at the International Centre for Theoretical Physics in Italy, first presented his theory of goyaks in 1986 and has continued developing this theory.

Goyaks are the title of this chapter—this look at physics beyond the standard model—because goyaks provide a prototype or paradigm for the radical theories of new physics that appear from time to time in publications of modern physics.

Goyaks have the additional cachet that, for a native speaker of English, the word goyak has a mysterious and original sound, and its translation shows its meaning to be elemental, basic, fundamental.

The theory of goyaks has many of the characteristics of the ten radical theories of new physics that we'll be discussing in Part Four. It "reveals the primordial deeper structures underlying fundamental concepts of contemporary physics."[293] Primordial, as an explanation that begins at the origin of the universe. Deeper, as it proposes a unifying conceptual structure for all of contemporary physics' particles and interactions (forces) as well as for the nature of spacetime itself.

Goyaks are the substance out of which both particles and geometry are made. Ter-Kazarian develops sophisticated mathematical formalisms for goyaks, and for reciprocal linkage between different types of goyaks, from which he builds the foundations of modern physics.

What goyaks share with other radical theories of new physics is a quest for a unifying structure that is aesthetically more appealing than modern physics' standard models, which require the acceptance without explanation of numerous apparently arbitrary assumptions.

And, beyond the scope of goyaks' radicalism, we'll also find, in our upcoming ten radical theories, theories that challenge the basics of quantum theory, and theories that introduce consciousness and the mind as integral elements of the physical world.

Goyaks also tell us something about the sociology of physicists and about the business of physics. It is natural that some toilers in a hidden world of nonconforming radical thinkers find no access to a world of Big Science. There is very little review, constructive criticism, or even destructive criticism of the toilers in this hidden world.

We'll discuss physics' standard model a bit more before exploring the sometimes private thinking of ten radical theorists of new physics.

***

The standard model of particle physics is the core of the academic physics curriculum and is not easily challenged. It is a comprehensive theory detailing the construction of matter and the interactions among particles. It unifies the electromagnetic and weak forces, into the electroweak force. And, through quantum electrodynamics (QED), it incorporates quantum physics into our understanding of this unified electroweak force. The standard model also, through quantum chromodynamics (QCD), incorporates quantum physics into our understanding of the strong force. And since QCD closely parallels QED, the standard model suggests, although does not yet fully confirm, a quantum Grand Unified Theory, unifying quantum physics with a unification of three of physics' four forces.

The standard model is enormously accurate, to a level of almost precise confirmation by all observations and experiments, at large scales and small scales, by astronomical observation of the cosmos and by submicroscopic observation of effects in particle accelerators.

So what's the problem?

Why don't we conclude that reductionism is now almost complete, and that following along the path of the standard model of particle physics will complete our understanding of all of physics, giving us our theory of everything by explaining all of the universe's elementary particles and every way in which they interact?

What Cracks Do Physicists Find in the Standard Model of Particle Physics?

One concern raised about the standard model is that the reconciliation of QED with QCD, needed for a Grand Unified Theory, is incomplete. That is, the electroweak force and the strong force are not fully unified.

In addition, the fourth force, gravity, presents a major problem for the standard model, which neither unifies gravity with the other forces nor presents an accepted quantum model of gravity.

The number of assumed quantities within the standard model—parameters that are inputs into the model, given to the model—is bothersome. The model assumes without explanation values for particles' masses and forces' strengths, and the model assumes that nature has given us three families (sizes) for each elementary particle. This is bothersome because it means that mysterious patterns are assumed, not explained. It makes one wonder if we're missing a deeper understanding in which these assumed quantities fall out naturally, rather than these assumptions being presented as arbitrary constants of nature.

Another concern about the standard model is the hierarchy problem and its bothersome inelegance. The vast jump in energies between the forces makes one wonder how such asymmetry can truly be nature's way.

The Higgs particle, hypothesized in the standard model to give mass to the particles carrying the weak force, as well as to other particles of physics, has never been observed, and to some is another troubling inelegance—hypothesizing a new particle or field because we have no other explanation for the creation of massiveness.

Baryon asymmetry is not satisfactorily explained by the standard model. Baryons are one type of matter particle (fermion)—the type of matter particle, including protons and neutrons, that feels the strong force which operates within atoms. Today, there is always a baryon/antibaryon matching: if a baryon is created or destroyed, its antiparticle (antibaryon) counterpart must simultaneously be created or destroyed. But baryons create matter, and there is matter in the universe today only because at or soon after the big bang there was asymmetric creation favoring baryons over antibaryons. The standard model's explanation is generally considered inadequate to account for the extent of baryon asymmetry in the observed universe.

There is now near-universal acceptance among physicists that the neutrino has mass, in contradiction to the standard model's assumption of a massless neutrino.

What Cracks Do Physicists Find in the Standard Model of Cosmology?

Closely aligned with the standard model of particle physics is a standard model of cosmology. Questions about the standard model of cosmology arise from observations of the galaxies.

Dark matter does not exist in the standard model.

Dark energy (also called the vacuum energy) does not exist in the standard model.

The standard model does not explain the origin of the initial conditions at the big bang, which create the specifics of our universe. Without an explanation, we're forced to rely on the arbitrariness of nature.

Will the Standard Models of Particle Physics and Cosmology Survive?

There are twenty-eight free parameters in the standard model of particle physics and ten free parameters in the standard model of cosmology.[294] Do our standard models truly give us the most basic understanding of the universe, if so many arbitrary constants of nature must be assumed, without inherent basis or meaning?

Some high-energy observations from both particle colliders and astrophysics cause physicists to scratch their heads about how the standard model can explain them. Experimental physicists observe phenomena that seem to extend beyond any formulation of the standard model, and theoretical physicists propose new physics structures that extend beyond the standard model.

Today's research literature is filled with observations that raise questions about, but may still be answered by, the standard model. In this vein—the vein of possible contradictions to the standard model, possible new physics signals—research papers discuss the exotica of jets and partons, excited leptons, leptoquarks, lepton-number-violating interactions, exotic fermions, and CP violation. They discuss spontaneous symmetry breaking, double beta decay, meson mixing, and virtual effects. They ask if there is new physics to be observed in black hole radiation, in the travel of gravitons along geodesics, or in the extent of cosmic air shower depletion.

If the Higgs mechanism doesn't exist, would it have to be replaced by new physics?

Is there a fourth generation (family) of elementary particles?

Does new physics appear in the stretching of DNA strands?[295]

Does an anomalous acceleration of the Pioneer 10 spacecraft require new physics for an explanation?

Do we need extensions of the standard model—new physics—to explain other astrophysical and cosmological observations, such as large-scale structure formation?

Can only new physics, and a move away from reductionism, incorporate consciousness into science?

The list goes on and on of subjects for which physicists ask: can only an extension of the standard model, can only new physics, explain these phenomena? The observation that fundamental physical constants may in fact vary over time. Exotic tau decay, charm and beauty flavor, Trojan penguins[296].

Nevertheless, in spite of this long list of questions about the standard model, physicists remain divided as to how severe are these shortcomings—whichever actually pan out to be confirmed shortcomings. Is what's needed just some tweakings of the standard model, or does the model need major renovation or even demolition and replacement?

Tweaks to the Standard Model

String theory, generalized to incorporate supersymmetry within the M Theory framework, is the most widely accepted extension of the standard model. But many additional frameworks are also under consideration.

Physicists look at general categories of supersymmetric extensions of the standard model, and they try to narrow down what are the minimum additional assumptions and constraints which may need to be added to the standard model in order to match physics' latest observations.

Even though there is no explicit confirmations of the Higgs particle or Higgs field, physicists have not given up on an explanation for mass deriving from Higgs phenomena. They consider revised models of the Higgs sector: composite Higgs, a light Higgs-like neutral boson, little Higgs models, two-Higgs-doublet minimal supersymmetric standard models.

Physicists consider possibilities for new strong forces, and for extra dimensions, large and compactified. They look at technicolor models and zevatrons[297]. They construct seesaw models to explain neutrino masses, neutrino left- and right-handedness, and the instability of matter.

Does the standard model need a tweak? Or do we need a more basic look at physics' strange phenomena, physics' phenomena beyond today's physical reach? Is there just too much of an accumulation of new, unexplained phenomena, outside of the straight-line directions of today's mainstream approaches to physics?

Strange Phenomena

Strangely, twentieth-century physics has taught us that gravity reshapes space. That matter and position aren't defined until we observe them. That space may have more than three dimensions. That particles can be instantaneously entangled. That we may have basic misunderstandings of mass and of time. That the universe may have superluminally inflated, that it is filled with dark matter and dark energy and black holes. That subatomic particles mysteriously tunnel into other particles' territory. That mathematics is just too effective as a tool to explain science. That art and fiction and religion may have serious scientific bases.

And twentieth-century physics has given us strange new ways to understand these phenomena: parallel universes, imaginary time, fractals and chaos and complexity, the physical basis of information, quantum gravity.

There seem to be a lot of phenomena here that are beyond our physical world and that can't be understood by today's physics.

**

New physics has given us many strange phenomena, as well as new and strange ways to understand these phenomena.

The development of physics during the twentieth century has flowed along two paths, general relativity and quantum physics. The struggle of twentieth-century physics has been to unite these two paths into one. And the main approach along these paths has been reductionism: slice and dice finer and finer, and eventually we'll master how every piece is put together.

But just because our history of science has taken us along two distinct paths, does this mean that physics' deepest understanding will come from uniting these paths, from linearly extrapolating to find where the line of general relativity intersects the line of quantum physics?

Or will the deep understanding of twenty-first-century physics come instead from an understanding of both general relativity and quantum physics that transcends both? Not the intersection of their linear extensions, but a broader framework from which they both—and much more—emerge.

When physicists today look at the results of their experiments and theorizing, many place their findings in a context of new physics.

How far should we stretch standard models, old physics, before the framework bursts wide open?

It's no surprise that largely underground radical thinkers, generally with little support or acceptance, propose some radical theories of new physics.

28.
Hidden Physics Countdown: Ten Radical Theories of New Physics

You've seen it on David Letterman. You've seen it on the Travel Channel and on countless other TV programs and in magazines. Now you'll see it in a book on quantum physics: a Top Ten list, this time of radical theories of new physics.

Perhaps you've also had the pleasure of traveling with Ulysses Press's *Hidden . . .* travel guides: *Hidden Mexico, Hidden Hawaii, Hidden Florida* and their other "adventurer's guides" that take the traveler to sites off the beaten track, little known and—undeservedly!—little noticed.

We will now combine these two popular phenomena to create a Top Ten list of the Hidden—ten little-known theories of radical physics, off the beaten track, but on the track to new physics and the mind.

***

We've been through a lot of new physics, as well as a lot of physics of the mind, some hidden, some not.

We've heard from Einstein and Bohr and Eccles and Stapp and Penrose and Zohar and many others.

Einstein

The theories of Albert Einstein barely belong in a discussion that will include physics' connection with consciousness, with the indeterminate, since Einstein was firmly a realist, for whom God does not play dice, and who was committed to causation and determinism. But it is precisely Einstein's firm realism that we must keep in mind in our countdown toward Hidden Radical Theory Number One: does a theory pass the test of reality? Not all of our Ten will pass this test, but some—such as frameworks in which the commitment is to demystify quantum physics and the quantum jump, to explain the quantum as classical—are highly aligned with realism. In fact, our Theory Number One—in spite of being shrouded in the most extravagant connections of physics with the mind, the universe, our brains, our biology, and our evolution—is a theory aimed at explaining the ordinary real phenomena of science and of our life experiences.

Perhaps Einstein would view Theory Number One as, in fact, realistic.

It takes a lot of twisting and irony to view Einstein as a contributor to theories of physics and the mind. To do so means that our best theories—no matter how ethereally formulated—must ultimately be firmly grounded in ordinary reality.

Perhaps a listmaker of the early twentieth century would have uncovered a Swiss patent officer's strange and radical theory of special relativity and listed it within an early 1900s' version of the Top Ten Hidden Radical Theories of New Physics.

Bohr

The Copenhagen Interpretation incorporates the theories of Danish physicist Niels Bohr and his colleagues. Bohr is truly the father of physics and the mind, which he advocated in opposition to Einstein and his realism.

The Copenhagen Interpretation of quantum physics is that quantum phenomena are not real until and unless they are observed. Indeterminism is a central concept of the Copenhagen Interpretation, in spite of this appearing to be a radically unscientific notion.

The debates originating with Einstein and Bohr have continued for many decades and are still unresolved. There is even today no consensus view among physicists of the role in physics for the observer, of the place in science for consciousness and the mind. But it is Bohr and the Copenhagen Interpretation that reserved a place for them at the table.

Eccles

You gotta love a psychon.

A paradigmatic emblem of mind physics is a drawing, appearing in Eccles' works, of three dendrons—grouped neurological dendrites— that are being invaded by floating hypothesized shapes, psychons, the purported elementary particles of thought.[298]

What visual imagery! Solid square psychons and open square psychons and solid circle psychons, floating down among dendron trees, connecting the mind and body through the characteristic unitary experiences of the brain's forty million dendrons linked with their psychons.

Wholly radical at its initial publication, Eccles' theory is ultimately reductionist in its picture of thought particles and its focus on the physiology of our nervous system for consciousness-linked quantum events of elementary particle scale. But Eccles' knowledge of both neurobiology and physics sets a stage for radical physicists' interests and beliefs in pursuing theories involving the mind and consciousness.

Stapp

Berkeley physicist Henry Stapp is one of today's most energetic advocates of the core role of the mind in physics, of Bohr meaning what he said, not proposing concepts that need to be excused or explained away. Stapp is a historian of mind physics and continues to be an important contributor.

Penrose

Roger Penrose brought great public prominence to theories of mind physics—in both scientific and general communities—with the publication in 1989 of *The Emperor's New Mind: Concerning Computers, Minds, and the Laws of Physics*. An original researcher of developments in quantum gravity, Penrose also has boldly promoted the connection between an understanding of the mind and physics' reconciliation between relativity and quantum physics.

Some specific mechanisms suggested by Penrose are, like Eccles', more tied to mainstream physics than they may at first appear, due to their basis in neurological quantum physics—due to neurological elementary particles' quantum events being the creator of consciousness and the connection between the brain and the mind. This, I believe, is a reductionist approach, not a holistic aproach: it is locating a site of quantum activity that creates the intersection of concrete physics with abstract consciousness. This contrasts with some of our Ten Hidden Theories, which draw from the accumulation of new physics phenomena a holistically revised world view.

To be fair to Penrose, *The Emperor's New Mind* is at its heart an argument against a prominent view of its time, Strong A. I., the strong view of artificial intelligence, the view that our minds are computers and that one day our computers will be minds. This is not today's most highly advocated view of the philosophy of consciousness.

Penrose is a Platonist: the mind, for Penrose, is reaching for connections to the universe's eternal truths, and this brings a whole additional layer of complexity and holism to Penrose's new mind, a complexity resonant with some of our Ten Hidden Theories. With his twistor theory, his models of tilings and brain plasticity, and his role for Gödel's theory that not all is provable, Penrose remains at the forefront of developments in both quantum gravity and the physics of the mind.

Zohar

Danah Zohar's work is holistic, not reductionist. In focusing on the new physics phenomenon of the Bose-Einstein condensate, Zohar finds a unifying mechanism for the phenomenon of consciousness.

Zohar, and David Bohm's implicate order within which much of Zohar's work resides, are firmly in the territory of new physics and the mind.

Hidden and Radical

Radical theories of physics continue to proliferate. A few gain prominence but most remain essentially hidden.

Many although not all of these radical theories of physics are theories of physics and the mind. The mind is not a stranger to quantum physics, due to the surprising role of measurement or observation in the basic quantum process of the collapse of the wave function. This process is so surprising that to many it is simply unbelievable. Even believers have a wide range of interpretations.

But the more startling connection between consciousness and physics would be a connection in the opposite direction: not how consciousness plays a role in physics, but how physics plays a role in consciousness.

Quantum physics' superposition of states seems to some to be similar to our brain's background process of thinking along multiple paths of possibilities. The reduction of the state vector (collapse of the wave function, or quantum jump) is then quantum physics' modeling of our mind's selection of which path it will follow.

Our conscious experience of the arrow of time may be a physical process of thermodynamic entropy increase linked to that other irreversible process, the collapse of the wave function.

Some place consciousness as a fundamental property of the universe and map the mind to cosmological processes. Others note physicists' affinity for symmetry, for mathematical elegance, for aesthetic simplicity, and ask: if the intangible, intelligent, subjective processes of consciousness play a role in the hard sciences, is this an indication of the oneness of mind and matter?

The Ten Radical Theories discussed in the next ten chapters have received little notice in either scientific or general review—disproportionate to their highly radical questioning of the nature of the physical world and the role of quantum physics.

With the 2002 publication of *A New Kind of Science*, physicist Stephen Wolfram has brought his radical mathematical modeling of a theory of everything well beyond classification as a hidden theory of radical physics. This development of a new complexity theory of physics will not appear in our Hidden Physics Countdown.

Nor will the work of the French twins Igor and Grichka Bogdanov find its way to the Countdown. Their theories, including ideas on the origin of the big bang, the nature of the smallest scales of space and time, and topology's role in physics, have created something of an uproar in academic physics. There seems to be a majority view among physicists commenting on the Bogdanovs' work that it is, at best, questionable physics. But many physicists just aren't sure. The work has been published in serious journals of physics, but some wonder if it's a combination of new physics jargon nonsensically streamed together, and others wonder if it made it into prominent scientific journals without adequate peer review. Some even wonder if it's a hoax, which would place it in the proud tradition of nerdly pranks. A number of commenters wonder if the Bogdanovs' work is comparable to the deliberate hoax perpetrated by New York University physicist Alan Sokal in 1996, who managed to have his paper "Transgressing the Boundaries: Toward a Transformative Hermeneutics of Quantum Gravity" published in a a social science journal. The more paranoid among physicists wonder if the Bogdanovs' work could be a delayed counterspoof by social scientists looking to get by physics journals' screeners, but in any case the incident has raised questions about academic peer review and even about the nature of modern physics. One thing that is clear is that the Bogdanovs' work fails the *hidden* screen for our countdown of radical theories of new physics.[299]

What we will see in our Hidden Physics Countdown of Ten Radical Theories of New Physics will include quantum physics brought into the classical world and into the world of the mind and consciousness. And we'll see radical reconceptualizations of the universe and consciousness's role.

Far and away the jackpot of this search—Radical Theory #1 of New Physics—is Pitkänen's topological geometrodynamics (TGD). This massively developed theory has it all—a physicist's model of the mind, along with every aspect of new physics that we've discussed, every phenomenon beyond the reach of today's physical world. And, importantly, quantum phenomena are understood in TGD to operate at every size (not just the submicroscopic). The mind is central to TGD physics. Parallel universes, extra dimensions, entanglement, tunneling, information—these are all concepts that fall out naturally from the framework of TGD.

The Search

Selecting just ten radical theories of new physics is quite limiting. Radical theories of new physics have been around just about as long as modern physics has. Contemporary physicists, proposing a unification of electromagnetism, general relativity, and quantum physics, for example, cite a 1903-4 proposal for the unification of electromagnetism and gravity as an earlier radical theory on the pathway to their own proposal. [300]

Who has traveled along this path of speculation? What better way to find out than to search every database to find publications that discuss all or many of the phenomena of new physics?

In seeking physicists' speculations on radical theories of new physics and the mind, my first searches of physics and other science databases requested articles and other publications that covered every one of these phenomena: quantum physics or quantum mechanics, and consciousness or mind, and extra dimensions, and entanglement, and tunneling, and gravity or quantum gravity, and just about every other combination of phrases from new physics.

And guess what? No single article had it all.

But as I relaxed the search, I was surprised to find that I didn't have to go too far before I found some high-correlation matches. This was a big surprise.

We find evanescent photons in the brain as the basis for a quantum theory of consciousness.[301] We find adaptive resonance permitting the mind/brain interface through quantum holography.[302] We look for the basis of physics in two-behavior physics, one dimensionable, the other nondimensionable.[303] We mathematically model, using category theory and the multiplicity principle, complex objects with multiple configurations that permit the emergence of atoms from elementary particles, and from which we observe the ultimate emergence of biological systems, neural systems, and consciousness.[304] We look at consciousness as a form of virtual organization,[305] and we look at quantum information in EEGs and in the brain[306]. We look at atoms and consciousness as complementary elements of reality,[307] and we look at the microworld's quantum fluctuations as deriving from an elementary consciousness in nature that may also be the source of consciousness in the brain and mind[308]. We look at psychomas[309] and quantum monads[310] as fundamental elements of consciousness, fundamental pockets of reality, which have mathematical and topological representations which permit extension to physics' forces, to the geometry of spacetime, and to other elements of frontier physics.

The wealth of serious discussion of new physics and the mind is astounding and overwhelming. Websites such as "Quantum Mind"[311]— sponsored by Consciousness Studies at the University of Arizona and the Intuition Network—offer quantum mind archives, discussion forums regarding quantum approaches to consciousness, and reports from quantum mind conferences held throughout the world.

Speculation

For the rest of this book, we will be speculating, based on a premise that linear extrapolation from twentieth-century understandings will not give us a framework for the twenty-first century. This speculation will derive both from a perspective within twentieth-century science, and a perspective outside of twentieth-century science.

From within twentieth-century science, linear extrapolation inevitably leads the investgation to quantum gravity, some form of gravity that works in both classical and quantum physics. Since only gravity—uniquely among the forces—reshapes space, one premise of our speculation is that we will not find a theory of everything by focusing on higher energies or smaller sizes. Gravity is different from the other forces. The other forces operate against a background of space and time. But gravity is spacetime. Gravity is special. And we can never understand it as an extension of the other forces. It's a different animal, and no phenomenon—even strings—can offer an explanation for both gravity and the other forces.

So from within the framework of twentieth-century physics, it is only the background-independent theories of gravity that have any hope of providing a twenty-first-century theory of everything. No theory in which gravity operates—like the other forces—against a backround of space and time can treat gravity with sufficient exceptionalism to explain it all. We need a different path.

Let's take a step back from the the framework of twentieth-century physics. Forget entirely the order in which we've made the discoveries of twentieth-century physics. Forget that general relativity was discovered independently of quantum physics. Forget the order in which we discovered the forces.

And when we forget the history of how we got where we are today, I believe that we are no longer driven to unite gravity with the other forces. And we are no longer driven to bring general relativity into the quantum world as our method of creating a theory of everything.

In fact, I believe a greater mystery is how to bring quantum physics into the classical world. This mystery is at the center of the earliest thoughts about quantum physics by one of its great founders, Niels Bohr: "it is decisive to recognize that, however far the phenomena transcend the scope of classical physical explanation, the account of all evidence must be expressed in classical terms."[312]

Some suggest that our model of the physical world must modify our current understanding of either general relativity or quantum physics. The approach of the majority of physicists is to modify—to demand the quantization of—general relativity. But a minority "argue that perhaps it is quantum mechanics that ought to be modified." Perhaps the preference for the more usual approach, quantizing general relativity, "comes in part from 'majority rule'—most interactions are very successfully described by quantum field theory, with general relativity standing alone outside the quantum framework—and in part from the fact that we already know a good deal about how to quantize a classical theory, but almost nothing about how to consistently change quantum theory".[313]

As he closes *Three Roads to Quantum Gravity*, Lee Smolin goes way out on a limb to make predictions about how our understanding of physics will change over the upcoming decades. His vision provides an excellent introduction to what we're about to discuss:

- "The present formulation of quantum theory will turn out to be not fundamental."[314]
- "The present quantum theory will first give way to relational quantum theory,"[315] according to which "the quantum state of a particle, or any subsystem of the universe, is defined, not absolutely, but only in a context created by the presence of an observer"[316] and there is "a division of the universe into a part containing the observer and a part containing that part of the universe from which the observer can receive information."[317]
- Eventually, the new unifying theory of physics "will be reformulated as a theory about the flow of information among events."[318]
- "By the end of the twenty-first century, the quantum theory of gravity will be taught to high-school students all around the world."[319]

***

Suppose we set out on the route of bringing the collapse of the wave function down to ordinary life, no longer keeping it on a pedestal of nonclassical exceptionalism.

I can't imagine this being accomplished without a key role in physics for consciousness and the mind.

And if this is where we start—first, bringing quantum physics into the classical world; second, bringing the mind into physics—how far can we go? Will this beginning take us to a framework in which new physics' strange phenomena beyond today's physical world simply fall out naturally from this framework? Will this journey—through physics and the mind, and through the whole constellation of new physics phenomena—take us to a new theory of everything?

This is the quest for Part Four. To search for physicists' radical theories, promulgated with little fanfare or notice, off the beaten track but on a track to new physics and the mind.

What the Top Ten theories share is a radical relook at twentieth-century physics. We'll find theories that are not tied to the twentieth century's historical ordering of discoveries in physics, theories with scope larger than the reconciliation of relativity with quantum physics, theories that are holistic, that do not presume that the road to deep understanding is by reducing our view to how the components of the components of the components work.

Fasten your seatbelts. We'll even go beyond TGD to the quantum universe and the question of what else is out there beyond our reach.

29.
Hidden Radical Theory #10: Olavo's Rederivation of Quantum as Classical

In a series of sixteen papers published from 1995 through 1997, physicist Leopoldino S. F. Olavo of the University of Brasilia took on the task of rederiving all of quantum physics as a classical theory. Olavo essentially brings the mysteries of quantum physics back to a Newtonian basis—classical physics, as built for two centuries on the foundation of Isaac Newton's eighteenth-century developments of modern science. Today, classical theory means physics up through special and general relativity, but not quantum physics.

Thus Olavo's theory, encompassing all of physics, is relativistic Newtonian physics—that is to say, a classical theory of quantum physics.

***

The Accepted View: Quantum Physics Corrects Classical Physics

Quantum physics, in the view of the vast majority of scientists, provides a more accurate view of the physical world than classical physics provides. Newtonian physics, as developed centuries ago, needs the corrections of relativity theory, but these corrections still keep physics in what is now called a classical view. Classical physics, in turn, has since been further corrected by quantum physics.

In addition to providing physics with significant additional accuracy, quantum theory has difficult implications, for which there is not a current consensus among scientists, about the nature of reality. The amplitudes of the Schrödinger equation for the quantum wave function tell us the probabilities for various outcomes of physics. But, under the Copenhagen Interpretation of quantum physics, no outcome actually occurs until an observation is made, at which point the wave function collapses to a randomly determined outcome, subject to the Schrödinger's equation's probabilities (likelihoods), but not actualized until observation.

Olavo: Classical Physics Is All There Is

Olavo turns the orthodox view on its head, viewing quantum physics as derivable from classical physics, and viewing quantum physics as an approximation of classical physics. His sixteen papers published from 1995 through 1997 present derivations, from classical physics, of all of the principal aspects of quantum physics.

Olavo's is a realist interpretation of physics. In Olavo's interpretation, the observer and the observer's consciousness play no role. No external psychophysical observers are postulated. Physics' descriptions of the world do not depend on mental activities. Measurements may be subject to statistical likelihoods and may be unknown, but they are not indeterminate, waiting for an observer.

Nor, in Olavo's construction, is the quantum a fundamental characteristic of the subatomic world. A core concept of quantum physics—the Heisenberg uncertainty principle—is not a fundamental property of nature, but instead is a consequence of limitations of quantum physics' adopted formalisms, its equations.

And in contrast to some other realist interpretations, which require a nonlocal hidden variable—information transmitted superluminally—Olavo's theory is local, with relativistic Newtonian physics being its only hidden variable, its foundation in reality. Olavo's theory is built on a totally deterministic and local theory, which is no more than Newtonian physics adjusted for relativity theory.

With these starting points, Olavo systematically presents algebraic derivations of all of quantum physics' operational formulas, including its most mysterious. He derives today's important quantum umbrella theory—quantum field theory—as a classical theory. He presents quantization—another core principle of quantum physics—and the quantum tunnel effect in purely classical terms. He describes diffraction and interference—thought to demonstrate the wave nature of the electromagnetic force's dual wave/particle essence—in terms of corpuscular theory, purely as a particle. He uses a concept of negative mass to explain irreversibility and the arrow of time. He explains spin—a property of elementary particles generally understood only in quantum terms—as a classical phenomenon related to the algebraic symmetry of his models. And he does all this without appealing to the usual quantum epistemology, such as observers and wave/particle duality.

Olavo's is a radical revision of quantum physics: revise quantum physics by getting rid of it, showing that it is superfluous and that it reduces to classical physics.

Olavo's work has not achieved broad review or recognition. Commentary in 2001 on Olavo's derivation of the Schrödinger equation demonstrated that Olavo's derivation applied only in narrow circumstances, not broadly.[320]

But as a hidden theory of radical new physics, Olavo's rederivation of quantum as classical receives an A for hidden and an A for radical, not to mention an A for effort.

Quantum theory, while extraordinarily successful in advancing science and technology, has also been extraodinarily bothersome in its indeterminateness and superposition of realities. Olavo is not the first to wish away quantum physics' bothersome characteristics as ultimately simply classical.

***

Other radical theorists share Olavo's discomfort with quantum quirks, in particular with the exceptional and interrupting event of the collapse of the quantum wave function. We will see more radical revisions of quantum physics as we proceed with our countdown, including theories for which, as for Olavo, classical physics is exact, rather than an approximation of quantum physics. However, in these other theories which we'll be discussing, quantum physics is also exact and is the deeper reality which classical physics, although exact, only represents.

When we reach Radical Theory #1 of New Physics, we will find a theory that provides a classical framework for quantum physics' configuration space of quantum possibilities: in TGD, our #1 Radical Theory of New Physics, this configuration space is an infinite-dimensional space of physically allowed spacetime surfaces, a world of classical worlds. TGD preserves only the quantum jump as physics' quantum, not classical, phenomenon. TGD embraces the quantum jump, physics' only quantum phenomenon, as a phenomenon operating at all length scales: in TGD, the quantum jump is a macroscopic phenomenon that we experience throughout our daily lives.

And TGD also embraces the mind. It is truly new physics and the mind.

30.
Hidden Radical Theory #9: Majerník's Geometric Description of Quantum Physics

Vladimír Majerník is a physicist at the Slovak Academy of Sciences in the Slovak Republic and at Palacký University in the Czech Republic. His 1996 article in the physics discussion journal *Physics Essays* proposes a geometric model of physics which explains quantum physics' central mystery, the quantum jump.

***

Majerník's work does not seem to have taken off as a widely referenced framework among his physicist colleagues. But it's an intriguing place for us to continue our speculations, in its bold attempt at demystifying quantum physics.

Majerník has a history of deep and questioning thinking about the nature of the quantum world. For example, in his 1969 article "Information Content in Quantum-Mechanical Systems," he links information to entropy and takes this linkage to a discussion of measurement of information in quantum systems. This role for information in physics has found a place in subsequent decades of modern physics.

In his 1996 article "A Geometrical Description of Quantum Mechanics," Majerník discusses the attractiveness to modern physics of three categories of models that incorporate extra dimensions beyond our familiar 3 + 1 (three spatial dimensions plus one time dimension).

We've talked a lot about one category of extradimensional models in which generally (although not always) the extra dimensions are compactified. Compactification explains why we're not sensitive to extra dimensions' existence. Experimental physicists seek confirmation of these extra dimensions, and theoretical physicists explain how even infinitely extended extra dimensions may go unnoticed.

In the second category of extradimensional models that Majerník discusses, time has three dimensions. Development of these models has largely been driven by a sense of symmetry between time and space: if space has three dimensions, why doesn't time? There has been a surprising amount of development of these models, since the 1950s, and three dimensions of time help answer open questions of physics, such as why the speed of light is constant, as well as the homogeneity question to which the cosmological model of inflation also provides an answer. But extradimensional models with three dimensions of time have not broadly taken off among physicists in the decades since they were first contemplated.

The third category of extradimensional models that Majerník discusses, which is the heart of his 1996 *Physics Essays* article, is a model in which there are four extra dimensions—three of space and one of time. But these extra dimensions are dimensions of nonreal spacetime, which exist in addition to the familiar four real spacetime dimensions.

Majerník's nonreal spacetime is based on complex numbers, which are numbers that have an imaginary component and may in addition have a real component. So we're back to imaginary numbers, which we've seen before with imaginary time. But here the extra dimensions of space too have imaginary components, and the extra dimensions of both space and time may in addition have real components.

Now generally imaginary numbers bother ordinary mortals a lot more than they bother physicists or mathematicians, who tend to view them as just like real numbers (only imaginary). But the fact is that imaginary numbers are integral to mathematical models of modern physics, so Majerník's extra dimensions of nonreal spacetime offer some intriguing food for thought as a geometric description of the quantum jump.

In Majerník's model, events occurring in nonreal spacetime are, at certain moments, projected onto the real spacetime in which we live. This can result in a quantum wave of possibilities, existing in the full eight-dimensional spacetime, being mapped through a quantum phenomenon onto a single realized event in real four-dimensional spacetime. This geometric projection of the eight-dimensional world's quantum wave onto our four-dimensional world is Majerník's geometric description of quantum physics.

Majerník brings the philosophical work of Plato into his discussion of the universe of four real dimensions plus four nonreal dimensions. We've heard from Plato earlier, in Penrose's work on the physics of consciousness and in Smolin's work on quantum gravity. Majerník cites Plato's conceptualization that there is a shadow space in which events are fully realized, but that the world we perceive consists of mere projections of these more fully realized events.

For Majerník, it is the eight-dimensional world in which the complexity of possibilities is fully expressed, a rich world of all that is possible. Our local four-dimensional world is the simple world in which quantum possibilities have been reduced to a single path.

Majerník applies his geometric description of quantum physics to explain several mysteries of theoretical physics, including the neutrino mass and quantum chromodynamics (QCD), the quantum theory of the strong force operating within the nucleus of the atom.

Majerník also develops a theory of bradyon-luxon symmetry, a theory similar to the more widely promulgated theory of supersymmetry. Supersymmetry (SUSY) proposes a sibling boson (force-carrying particle) for each fermion (matter particle), and a fermion for each boson. In Majerník's bradyon-luxon symmetry, photons and other massless particles move at the speed of light in real spacetime, as in our customary understanding. But in Majerník's complex (nonreal) spacetime, it is massive particles that move at the speed of light, and photons and other massless particles that travel subluminally and can be localized. And it is the projection of wave processes in complex spacetime onto real spacetime that permits Majerník to create a geometric model of quantum physics.

***

Majerník's work proposes a solution to a central mystery: can we develop a classical understanding of quantum physics? In addition, Majerník applies his theory to explain problems of new physics.

Majerník's geometric description of quantum physics has a number of similarities to TGD—topological geometrodynamics—which we will be discussing more thoroughly.

TGD's universe is also eight-dimensional, and real four-dimensional physical spacetime is also the projection of physically allowed quantum possibilities. In TGD, though, four-dimensional bounded universes encompass and create the spacetime for smaller four-dimensional universes, in many-sheeted spacetime. Each bounded universe is four-dimensional but treated like a point within in a larger four-dimensional universe. Our bodies, for example, are four-dimensional universes. So are our organs, each also four-dimensional universes, operating within the larger four-dimensional universe of our bodies. So each organ's four dimensions are understood in TGD within the encompassing body's four dimensions. TGD permits both of these (and many more) four-dimensional universes to coexist by connecting the embedded universes through tiny four-dimensional wormholes, thus placing four-dimensional spacetime sheets within eight-dimensional embedding space.

TGD also brings the mind and consciousness into the heart of physics, through the quantum jump, which in TGD is physics' only true quantum phenomenon. And TGD also has a central role for a special mathematics called p-adic mathematics, which has been used for several decades to model thought and memories. In TGD, the real world's many-sheeted spacetime is also understood p-adically, through p-adic mathematics, which is the mathematics of enclosure and which is the natural mathematics of cognition.

Let's continue pressing on toward Radical Theory #1, introducing more and more elements of new physics and the mind.

31.
Hidden Radical Theory #8: Avery's Dimensional Correspondence

Like Vladimír Majerník, Samuel Avery has published in *Physics Essays* a radical resolution to core issues of modern physics.

Avery's 1990 discussion looks at current mainstream approaches to physics' core problem of the reconciliation of relativity with quantum physics, and he is concerned that many physicists have feared treading outside the "bounds of science" into the territory of mental processes, even though they know that physics' strange phenomena at "dimensional extremes" have to do with the process of observation. Examples are the quantum jump occurring at the smallest scales, and the time dilation, length contraction, and higher masses that occur at the highest speeds (near the speed of light).

Avery's analysis has led him to a radical theory connecting matter with perceptual consciousness: mass becomes a second dimension of time, and the theory's five dimensions of spacetime are connected to our five senses.

Noting that he's discarded the concept of mass "merely" in order to reconcile relativity theory with quantum physics, Avery asks: "Is it worth replacing an entire world view for the sake of a few strange observations in particle accelerators and interstellar thought experiments?"[321]

His answer, of course, is yes.

***

In Avery's theory, dimensions are structures of consciousness, structures arranged to create the appearance of matter. For Avery, modern physics is best understood if we move away from a "materialist world view" to a concept of matter as a fifth dimension of spacetime.

In particular, this fifth dimension is a timelike dimension, a second dimension of time.

This connection of mass with time comes from thinking about both inertial mass (which relates acceleration to force) and gravitational mass (which measures the strength of an object's gravitational pull). When we measure either acceleration or gravity, the units of our measurement are units of length per unit of time per unit of time (for example, meter per second per second). It is the second of these two units of time that, in Avery's model, represents gravity's reshaping of four-dimensional spacetime, and that is best considered a fifth dimension of spacetime and a second dimension of time.

With five dimensions, Avery is now ready to create "dimensional correspondence" with our five senses.

The realm of consciousness in which our sense of touch operates is the realm of the second time dimension, which is also the dimension of mass. Thus tactile sensations are similar to the sensations we feel under the influence of gravitation or of acceleration, sensations dependent on a second dimension of time.

Avery goes on to align the other four senses with the standard four dimensions of spacetime:

> taste: one temporal dimension
> olfactory: one spatial dimension
> auditory: two spatial dimensions
> vision: three spatial dimensions

For example, a single olfactory observation can tell us distance, but not direction. But an auditory observation can also tell us direction. Vision adds depth. But by contrast taste and touch require us to be in contact with what we're sensing and therefore do not provide spatial information.

Once the basics of dimensional correspondence are mastered, the most robust concepts of the theory—the ones that solve physics' deep mysteries—are the concepts of dimensional interchange (among spatial and temporal dimensions) and tactile reduction (reducing the more complex senses to their tactile elements). This is how Avery brings dimensional consciousness to the quantum level. And, operating in the opposite direction, Avery sees here the evolutionary development of multicellular consciousness from experiences of the single cell.

Avery does confess that the theory is "difficult to conceive initially",[322] and few would doubt this. But what I find particularly interesting is Avery's motivations in constructing this theory, his measures of success.

For Avery, one measure of success for dimensional correspondence is that, compared to mainstream theories, it explains more experimental results with fewer assumptions. Mass ("the materialistic world view") is unnecessary. And the speed of light—which has a constant value, and which is the universe's greatest speed—is reconfigured as the connection between spatial perceptual consciousness and temporal perceptual consciousness, rather than being a speed within spacetime.

A second of Avery's measures of success is that he links mind and matter—he creates "evidence of a physical continuity between perceptual consciousness and material substance."[323]

In addition, Avery is not satisfied with a mere "operational theory", one that "promotes specialized or applied research at the expense of the intellectual expansion that is the real fruit of scientific inquiry."[324]

And finally, Avery sets out not only to solve mysteries of new physics, but to do this in a way that is consistent with our ordinary experiences. He is not aiming at explanations that are meaningful only at dimensional extremes; he's aiming at a new theory of the ordinary that also explains the extraordinary.

Whatever view one has of Avery's achievements according to his measures of success, these measures themselves do stack up well with where we're headed with our speculations. Our most desirable theories of new physics and the mind are those that exhibit economy of concepts, that derive from connections between mind and matter, that expand physics' intellectual framework to naturally incorporate the exceptional phenomena of new physics, that are grounded in real experiences.

In a 1992 follow-up *Physics Essays* article, Avery responds to comments he received on his earlier article and explores in particular the role of the observer. Avery notes parallels between the indeterminancy of observers and the quantum indeterminancy of subatomic particles and of energy quanta. This mapping between mind phenomena and physical phenomena, and the assertion that both realms are understandable as quantum phenomena, will recur as we explore additional radical theories of new physics.

***

Avery's goals set a stage for much that follows in our countdown. But in spite of dimensional correspondence's grand goal of linking physics with our consciousness of perception, Avery's proposal can be seen as as a reductionist linkage—directly assigning dimensional configurations to our senses—rather than one that proposes a rethought holism for the physical and conscious worlds.

32.
Hidden Radical Theory #7: Botta Cantcheff's Phenomenological Spacetime

Brazilian physicist Marcelo Botta Cantcheff has investigated various routes to the unification of physics' forces and to the creation of an understanding of quantum gravity. These routes include topology, generalized geometries, arbitrary numbers of extra dimensions, and other mathematical models and variations on field theories.

We discuss here Botta Cantcheff's phenomenological spacetime—his derivation of quantum gravity from a definition of spacetime in relation to the phenomenology of physical interactions.

***

Botta Cantcheff looks at the usual meaning of spacetime and notes that, for a specific event—a specific physical system or observation—only a subset of spacetime has effective physical significance. This is the subset of spacetime local to the event, the subset of spacetime that can be in causal contact with the event because it is within the reach of communication that travels no faster than the speed of light. The rest of spacetime is nonlocal and is therefore without effective physical significance.

Phenomenological spacetime is not in existence before a physical event occurs, but instead develops together with the event. The causal structure of spacetime corresponds to the causal connections among the phenomena that create a physical event.

The concept that phenomena create the structure of spacetime, rather than take place within or against a backdrop of already-structured spacetime, results in the background independence that reflects the gravitational exceptionalism—only gravity reshapes space—critical to the most satisfying formulations for quantum gravity.

Botta Cantcheff goes on to examine physical models of motion, the simplest of which are those for a single particle. He notes similarities between physics' mathematical modeling of the motion of a single particle and the mathematical modeling of matter fields in physics' field theory, and comments: "We claim that this is not a simple coincidence, but it rather reveals a fundamental fact of Nature with strong consequences."[325] These consequences are the creation, from one-particle models of motion, of a unified field theory for physics' matter particles and physics' forces, including the force of gravity.

Botta Cantcheff accomplishes this unification by creating his unified field theory in analogy with one-particle physics. A single particle, moving along its geodesic—its most efficient path within its four-dimensional spacetime—represents the dimensional reduction of the higher-dimensional space of matter fields in which physical spacetime is embedded. The physical spacetime paramaterizes the evolution of M-space, the space of matter fields.

This FT-OPT Theory—this universal correspondence between Field Theory and One-Particle Theory—allows Botta Cantcheff to extend the quantum physics of a single particle to a full quantum field theory, uniting quantum physics with general relativity's reshaping of spacetime, and uniting physics' forces and matter particles.

This unification is to be understood in the context of phenomenological spacetime, in which phenomena create spacetime as they proceed. In Botta Cantcheff's terminology, he thus creates *by construction* a background-independent quantum theory of gravity.[326]

***

With Botta Cantcheff's phenomenological spacetime, we continue to foreshadow elements of our later theories—many-sheeted spacetime as Botta Cantcheff's locally created universes within larger nonlocal spacetime, the quantum jump's continual recreation of superposed physical spacetime possibilities as Botta Cantcheff's continual phenomenological recreation of spacetime.

In our next theory, local topologies appear throughout a spacetime lattice, and an event in a higher-dimensional space corresponds to many three-dimensional topological spaces—to all of those topological spaces that could, by quantum theory, possibly localize the higher-dimensional event.

33.
Hidden Radical Theory #6: Spaans' Topological Dynamics

During his years at the Kapteyn Astronomical Institute in the Netherlands, at the Harvard-Smithsonian Center for Astrophysics, and at Johns Hopkins University, Marco Spaans has authored or co-authored numerous articles on the structure and chemical composition of the universe. He also has written a series of articles over the years on the topological properties of spacetime, which lead him to far-reaching conclusions about general relativity and quantum physics.

***

Spaans has for many years researched the topology of spacetime—the shaping of the dimensions of the universe. Spaans' work—his Topological Dynamics—has allowed him to draw conclusions about the cosmological constant, which measures the universe's expansion characteristics, about black hole entropy, about the cosmic microwave radiation and its fluctuations, about dark matter, about the mass-inducing Higgs field and the equivalence of gravitational and inertial mass, about superstrings and M Theory, and about the mass hierarchy of the elementary particles.[327]

Spaans also focuses his attention of the topology of Planckian spacetime—quantum spacetime, space and time at their theoretically smallest magnitudes. It is at this level—the quantum foam—that gravity is unified with the other forces, and that general relativity is reconciled with quantum theory.[328]

Topological Dynamics is Spaans' proposal for the underlying physical principle that unites general relativity and quantum physics.[329] In Topological Dynamics, the universe has a ground state and an excited state. The topological manifold—the geometric arena—of the universal ground state is a specific topological structure, which Spaans labels Q, with twenty-three degrees of freedom, similar to twenty-three dimensions. Spaans demonstrates that this ground state manifold, Q, is not an arbitrary result of nature but instead can be derived from basic principles.

The universe's excited state resembles a lattice of three-dimensional structures. L(T3) is how Spaans mathematically labels this underlying structure of Q. The excited state is the universe's direct reflection of the Heisenberg uncertainty principle, providing the boundaries on the physical realities we experience, and also suggesting an interpretation of quantum entanglement's instantaneous linking.

In Topological Dynamics, spacetime "provides both the stage for and the performers of physical phenomena."[330] The topology and geometry of spacetime accomplish this by the interplay between the topological manifold of the universal ground state and the the excited states' lattice of three-dimensional topologies.

The Heisenberg uncertainty principle tells us that there are limits to how precisely the position of a particle can be localized. Nature must therefore allow, at every moment in time, a complete set of three-dimensional spaces, permitting every possible localization, to be constructed within a linked lattice.

In the mathematics of topology, size doesn't matter. Only shape matters. And the ground state manifold Q is built up from two shapes—handles and three-tori. The shape of a handle is just what you'd think: a structure of any size, with an open hole in it, attached to another structure in a way that preserves its hole. A torus—pluralized as tori—is a doughnut shape, again with any sizing of its circumference, thickness, and hole. And a torus does not even have to have a rounded shape, as long as it preserves the topological characteristic of a connected outer surface with a hole in it. A three-torus is a three-dimensional torus.

Spaans brings gravity into Topological Dynamics through the handles, and he brings quantum physics—the superposition principle and quantum field theory—into Topological Dynamics through the three-tori. He then assumes three quantities as givens: the speed of light, the Planck length, and the Planck mass. From this topological structure, and with only these three unconstrained numbers, Topological Dynamics predicts the masses and other important quantities of physics' standard model, explains the excess of matter over antimatter that is one of modern physics' unexplained phenomena, and incorporates a model for dark energy and for other important phenomena of astrophysics and particle physics.

Implications

Spaans shows that topological analysis—looking at shapes but not sizes—can result in conclusions of enormous scope.

Spaans suggests that Nature's three unconstrained quantities only "fix the scale of physical phenomena, but are not relevant for the nature of the underlying physical system." He suggests further "that Nature does not even distinguish physical reality on this level, even though our perception of phenomena depends strongly on the numerical values these three parameters obtain."

This gives us a proposal in which "Nature distinguishes all phenomena through geometric and topological information." [331]

***

As we go on with our countdown of hidden radical theories of new physics, we will find physicists who do not deny the importance of topology, but who also look to understand even Spaans' three unconstrained quantities as deriving from a more elemental reconceptualization of the nature of the physical world. And we'll also be introducing physical models of consciousness and the mind, which are not emphasized in Topological Dynamics but which are the special emphasis of our next theory.

34.
Hidden Radical Theory #5: Kafatos and Nadeau's Conscious Universe

Menas Kafatos and Robert Nadeau of George Mason University discuss *The Conscious Universe: Parts and Wholes in Physical Reality* in their 1990 book, updated to a second edition in 2000. They look at human consciousness—"self-reflective awareness founded on a sense of internal consistency or order"[332]—and, provocatively, extend this sense of consciousness to the undivided wholeness, the cosmic order, of the universe.

***

Complementarity
Complementarity is a core concept of quantum physics, according to which certain physical properties are innately paired in such a way that knowing more about one property means we will know less about the second property. This concept is quantified in the Heisenberg uncertainty principle, for which two frequent applications are the position/momentum conjugate pair and the energy/time conjugate pair. The more precisely we can quantify a quantum entity's position, the less precisely we can quantify its conjugate attribute, momentum. Sometimes this is simplified as the more we know about where a particle is, the less we know about where it is going. Measurements involving energy and time show this same phenomenon of complementarity.

Complementarity is a concept central to the Copenhagen Interpretation of Niels Bohr and other quantum physicists who focus on the dualism of complementary concepts. Constructs such as wave and particle may appear classically incompatible, but in quantum physics they unite as a dualism, two constructs that only together describe reality, even if only one applies at any single instant.

The Schrödinger equation for the quantum wave function describes, precisely and deterministically, the evolution of a quantum system, as long as the system is not observed or measured. Unobserved and unmeasured, this is a massively parallel evolution, with all possible evolutions superposed, unfolding in parallel.

Upon observation or measurement, the world is no longer deterministic, it is probabalistic. These probabilities are given to us as an additional application of the Schrödinger equation. A particle will be observed or measured not at a specific location determined in advance, but at a location for which we can know in advance only what is likely and what is unlikely.

For Bohr, and for Kafatos and Nadeau, both sides of this create a seamless reality—both the determining wave and the determined particle. We cannot separate out either the wave or its collapse to a particle as more real than the other.

The New Epistemology of Science

Kafatos and Nadeau take this further. They apply the complementarity principle to view the quantum jump as the dual (complementary) phenomenon to deterministic quantum rollforward. In this view, the quantum jump is not a special or privileged quantum event; the quantum jump is the conjugate attribute of deterministic quantum evolution, and these two conjugate attributes can be understood only as dual aspects of a quantum whole. This results in a central role for consciousness in Kafatos and Nadeau's physics within the atom and within the universe.

Kafatos and Nadeau view their theory as creating a new epistemology of science, a new understanding of the nature of scientific knowledge. Kafatos and Nadeau's epistemology of science is holistic and irreducible, with consciousness universal. They apply this epistemology to biology, to the emergence of living systems, and to the nature of evolution. They also apply it to cosmology, the emergence and unfolding of the universe.

Kafatos and Nadeau note linkages between the macroscopic and micropscopic universes, and they note the extraordinary extent of the fine tuning of the universe. They bring in biological or organic life as a complementary construct to the physics of matter and forces. Consciousness is embedded in the universe, at all levels, at all stages and scales.

A New Stage of Consciousness

Kafatos and Nadeau extend their theory to propose an alternative to cosmological inflation and to even the big bang. Beyond this, they also propose a "new dialog between science and religion"[333] and wonder if our emerged knowledge and understanding of consciousness and the universe have brought us to a new stage in the evolution of consciousness. As we apprehend the "single significant whole"[334]—the dualism of consciousness and the physical universe—do we not only advance science, but also bring human consciousness to its next, more advanced stage?

***

Kafatos and Nadeau's elegant framework is at the far extreme from the reductionism that has been physics' major trend for decades. Their holistic framework has many similarities to Pitkänen's, discussed in Chapter 38 as our Radical Theory #1 of New Physics, especially in their shared holistic focus and shared view of quantum physics acting at all scales, not just the smallest scales.

But Kafatos and Nadeau's application of the complementarity principle to explain the collapse of the wave function as the dual phenomenon to determinisitc Schrödinger time evolution is a different stance than Pitkänen's on a central question of quantum physics. This central question is how to incorporate deterministic Schrödinger and the quantum jump into a single theory.

Pitkänen, rather than appealing to complementarity to de-privilege the quantum jump, embraces the quantum jump as the moment of consciousness, a special moment of leap from one deterministic time evolution to another. Thus privileged, Pitkänen's quantum jump permits a special place in physics for consciousness, and permits his own development—through topological geometrodynamics—of the linkages between our minds and our bodies, and between our biology and the universe.

35.
Hidden Radical Theory #4: Sirag's Reflection Space

The 1994 conference "Toward a Scientific Basis for Consciousness," held at the University of Arizona, included discussions of consciousness from the perspectives of psychology and cognitive science, the philosophy of mind, neuroscience and other medical and biological sciences, neural networks, models of consciousness, and quantum theory and nonlocal space and time.

We've already heard from a number of the presenters at this conference—Zohar on consciousness and Bose-Einstein condensates,[335] Wolf on the emergence of self-awareness,[336] Penrose and his frequent collaborator Hameroff on brain microtubules.[337]

The conference's papers on nonlocal space and time offered perspectives on new physics and the mind. One framework presented is a dualist mind/body model in which the mind exists in hyperspacetime, outside of normal spacetime.[338] Another framework refers to psychologist Carl Jung's concept of synchronicity—acausal transmission, faster than the speed of light, including connections between psychic and physical events—which Jung relates to the activation of archetypes within the collective unconscious. Jung's concepts are brought into a framework of physics and the mind through mechanisms of emergence of nonlocal information in the brain and through connnections between the mind and the collapse of the quantum wave function.[339]

Also as an application of nonlocal space and time, Saul-Paul Sirag develops a mathematical strategy—using the mathematics of "reflection space"—for a theory of consciousness.[340]

***

Reflection space is an extradimensional space with the special property that crystal structures within reflection space are reflected back upon themselves. These hyperspaces, and the hyperdimensional crystallographic structures residing in them, have been studied mathematically since the 1930s.

Sirag uses the mathematics of reflection space in models of particle physics, and he notes its use by others in numerous applications of physics, including optics, wave mechanics, gravity, the Heisenberg uncertainty principle, and twistor structures, which are at the heart of physicist Roger Penrose's approach to quantum gravity.

Sirag also examines specific reflection spaces in the context of superstring theory, and he identifies a particular reflection space—labeled the E7 reflection space—which has characteristics permitting the superstring modeling of all forces and matter in eleven-dimensional spacetime. This E7 reflection space also has applications in coding theory, including error-correcting code, and more generally has applications in analog-to-digital data transformations.

Sirag models the nodes and connecting lines of crystallographic structures in E7 reflection space, where—as in all reflection spaces—reflections map one crystal vertex into another. It is here that Sirag introduces both the quantum jump and consciousness: each reflection changing one vertex into another is a fundamental process of consciousness, an instant of awareness. Our nervous system presents us with a coherent means to experience these minute "blips" of awareness at each moment of time.

Sirag's reflection space model is a model of hierarchical embeddings of lower-dimensional structures within higher-dimensional structures. Higher-dimensional space is projected onto fewer dimensions, and our minds reflect shadows of more complex geometry.[341]

Thus, in Sirag's reflection space, our experience of consciousness is the mathematical projection of a more complex, higher-dimensional world onto the simpler spacetime in which our biology permits us to operate.

***

In Sirag's reflection space, consciousness is a projection phenomenon, a mathematical reflection within extradimensional physical spaces.

This isn't quite the extra dimensions of topological geometro-dynamics, our #1 Radical Theory of New Physics. In TGD's many-sheeted spacetime, eight-dimensional spacetime is the embedding space of multi-sheeted spacetime's four-dimensional universes. TGD also mathematicizes consciousness—through a special mathematical system called p-adic mathematics, which others have also used to model thought and the mind, and which TGD formally connects with the physical world.

p-Adic mathematics and thought space are, respectively, at the hearts of our #3 and #2 Radical Theories of New Physics, which we'll discuss on our way to TGD, our final destination.

36.
Hidden Radical Theory #3: Sidharth's Quantum Black Holes

B. G. Sidharth, of the Centre for Applicable Mathematics and Computer Sciences at India's B. M. Birla Science Centre, is a prolific author and researcher on topics of modern physics.

Sidharth has hypothesized about the fractal structure of the universe,[342] about holistic cosmology,[343] and about dualities and linkages between the microworld and the macro-universe.[344]

He has studied Earth's magnetic field—geomagnetism—and looked to temperature effects on electrons in Earth's solid core to explain puzzling reversals in our planet's magnetic field.[345] He has studied the magnetic field about Jupiter[346] and astronomical links connecting early civilizations of the Indus Valley in India with Mayan Central America and Easter Island.[347] He has studied the origin of life on Earth, and extraterrestrial chemical sources for some of life's ingredients.[348]

Sidharth notes two recent important discoveries of physics—the confirmation that neutrinos have nonzero mass, and the confirmation that the universe is ever-expanding, perhaps at an accelerating pace—and he discusses how these discoveries create the need for new physics models beyong the big bang and beyond the standard model of particle physics.[349]

Quantum black holes are a central concept of Sidharth's analyses, leading to new frameworks for understanding particle physics and astrophysics, for unifying physics' forces, and for reconciling quantum physics and general relativity.

***

Quantized Fractal Spacetime

Sidharth has explored the implications of quantized spacetime—space and time being granular, not continuous, at its smallest scales. He discusses how the quantization of time—the existence of the *chronon* as time's smallest quantum—leads to an explanation of the arrow of time.[350] And quantized spacetime leads to extended particles—elementary particle building blocks that aren't just dimensionless points, but that have extended dimensions, as string theory postulates.

Quantized spacetime is fuzzy spacetime, exhibiting the fractional dimensionality of complexity theory's fractal geometry. This leads to quantized fractal spacetime, a framework in which Sidharth places, and expands upon, quantum superstring theory.[351]

Quantized fractal spacetime's geometry is not our familiar geometry, the geometry dating back to Archimedes and other ancients. Quantized fractal spacetime is non-Archimedean, or ultrametric: lengths and distances cannot be measured as in our familiar Archimedean geometry. Quantized fractal spacetime is also noncommutative: its geometry, its spacetime, is not flat or ordered according to our usual formulas of geometry and algebra.

Sidharth uses a different form of mathematics—*p*-adic mathematics—to describe the physics of his quantized fractal spacetime, his fuzzy noncommutative spacetime.[352] We'll save the details of *p*-adic mathematics for Chapter 38, where we'll be discussing Pitkänen's extensive and central use of *p*-adic mathematics within topological geometrodynamics. For now, we'll just mention that other physicists, too, have found applications of *p*-adic mathematics to their work, and—as we'll be detailing further in Chapter 38—mathematicians have demonstrated that *p*-adic numbers have a unique place alongside real numbers as the only two complete mathematical systems.

Sidharth refers to his quantized fractal spacetime, his fuzzy noncommutative spacetime, as "a new paradigm for a new century."[353] It is a paradigm that explains some of physics' longstanding mysteries, such as how the elusive magnetic monopole comes to exist and why it is so hard to detect.[354] And much more generally, through quantized fractal spacetime Sidharth has proposed a unification of electromagnetism with gravity,[355] and in addition brought physics' strong force into this unification.[356]

Quantum Black Holes

Quantum black holes are a central mechanism for Sidharth's

unification of physics' forces, and ultimately for a unification of quantum physics and general relativity. Sidharth's quantum black holes are based on the Kerr-Newman black hole that would result from the collapse of a spinning massive star, not the earlier-modeled Schwarzschild black hole, which derives from the collapse of a stationary star.

Sidharth identifies matter particles—fermions—with quantum black holes. This is, arguably, the key to all of Sidharth's theories. The horizon of the black hole—the dividing surface from inside of which nothing, not even light, will ever overcome the black hole's gravitational force—corresponds to a critical quantum wavelength, called the Compton wavelength. The Compton wavelength is based on the work for which Arthur Compton won a share of the 1927 Nobel Prize for physics, for his demonstration of X-rays' dual wave/particle nature. In Sidharth's fractal universe, Sidharth characterizes the Compton wavelength as comparable to the thickness of the brushstrokes with which all of Nature is painted. And quantized spacetime solves a problem that must be addressed in models that incorporate black holes, preventing the model from reducing to the naked singularity that appears deep inside a black hole.[357]

All of Sidharth's modeling—quantized fractal spacetime, fuzzy spacetime, ultrametric and noncommutative mathematics—is brought to bear here. As he moves away from conventional spacetime, Sidharth generalizes the Heisenberg Uncertainty Principle[358] and challenges conventional notions of scale. He looks at recent experimental suggestions of variation in the fine structure constant, which determines the smallest relevant quanta of space and time,[359] and concludes that dimensionality is not absolute, but depends critically on the scale of resolution, from the Planck scale to cosmological distances.[360] And then the next leap: to quantum effects at multiple macro scales, universally.[361]

Fluctuational Cosmology

The logical result of Sidharth's thinking is a framework of *fluctuational cosmology*. In this framework, the universe was created as a phase transition, a fluctuation in the background zero-point field.[362] Dark energy—the mysterious force fighting gravity, pulling the universe outward—is one consequence.[363] And more generally, fluctuational cosmology describes the emergence—from a chaotic universe at the Planck scale—of quantized spacetime, the cosmology we inhabit, and all of the laws of physics.[364]

***

Sidharth's work, summarized only briefly here, is of great intellectual breadth and depth. His applications of p-adic mathematics will be reflected in our #1 Hidden Radical Theory of New Physics, which intertwines the real physical world and the p-adic world of the mind.

First, though, our #2 theory, a theory of thought space, the world of the mind.

37.
Hidden Radical Theory #2: Samal's Thought Space

Manoj K. Samal, of the S. N. Bose National Centre for Basic Sciences in India, has written or co-written papers on tunneling, gravity, topology, optics, nonlocality in classical physics, and other theoretical discussions of the interface between quantum and classical physics.

Samal also has a particular interest in science's role with respect to consciousness and the mind, discussed in this chapter.

**

"Can Science 'Explain' Consciousness?" is the title of an essay Samal presented as an overview of the current state of scientists' thinking about consciousness.[365] Samal points out that there are several aspects or components of consciousness:

> reflection—"the recognition by the thinking subject of its own actions and mental states"
> perception—the "faculty of being mentally aware of external environment"
> volition—"free will"

But Samal points out further that at its heart consciousness is an integrated, holistic experience.

Samal discusses views of consciousness from physics, neurobiology, and artificial intelligence. He is quite skeptical that any reductionist paradigm will succeed at providing a deep understanding of consciousness. For example, with respect to neurobiological approaches, Samal discusses the application of modern scanning technology in an attempt to locate functions of consciousness at specific regions of the brain: "Can this paradigm of finding neural correlates of the attributes of consciousness be fruitful in demystifying consciousness? Certainly not! . . . The currently prevalent reductionist approaches are unlikely to reveal the basis of such holistic phenomenon as consciousness."[366]

Similarly, Samal cites Penrose's discussion of Gödel's incompleteness theorem to argue for the limits of artificial intelligence's prospects for truly replicating consciousness.

And with respect to physics, Samal looks for holistic perspectives and is intrigued by the possibility of consciousness being related to the quantum vacuum, from which the universe emerged as a quantum burp at the big bang, and from which virtual particles continually emerge for an instant as part of ongoing physical phenomena. Samal offers *substantial nothingness*—reminiscent of Genesis's *astonishing emptiness* that we discussed in Chapter 25—as a synonymous term to quantum vacuum.

Samal also discusses the possible role of entropy as a physical basis for defining consciousness in terms of a system's degree of self-organization or complexity.

Ultimately, Samal aims at *quantum communicability* as a central concept for how consciousness may manifest itself scientifically, using the quantum superposition principle to explain the fuzziness—a term from complexity theory—of our experience of consciousness.

Thought space or T-space or mind space is how Samal begins to formalize his framework for a unified theory of matter and mind.[367]

Samal puts forth that consciousness is not an accidental phenomenon of the universe, but rather is a "fundamental property that emerges as a natural consequence of laws of nature."[368] He looks to a trend of quantum physics in which information plays a more fundamental role than even matter or energy. And from here it is not a big leap to a central role in physics for consciousness, the facility for processing information.

We've all noted animals' natural abilities to probe spectrums beyond human capabilities—dogs into ultrasound regions, bats sensing prey with echo techniques. But look at the range of experience humans can now probe with artificial aids of our own making: Samal cites a range of almost a trillion trillion trillion trillion (10^{47}) orders of magnitude, from 10^{-17} centimeters in particle accelerators to galactic superclusters of 10^{30} centimeters. And our explorations of space, matter, and energy are ultimately explorations through our minds.

Consciousness can be reduced to information, says Samal. And it is through information that we will map thought space, a topology obeying the laws of complexity theory.

In the philosophies of India, three states of sense awareness are noted: the conscious awake state, the subconscious dream state, and the unconscious state of dreamless sleep. And in Indian philosophy a fourth, superconscious state of mind transcends the limitations and constraints of (3 + 1)-spacetime. This *turiya* state is none of the other three states but is a combination of all of these three states. The techniques of Indian philosophy's unification of mind and matter guide practitioners towards accessing this state of *turiya*, this thoughtless but awake state, this state beyond (3 + 1)-spacetime.

Samal proposes a "world matrix" in which the physical and biological worlds operate within the manifold—or geometric arena— of traditional (3 + 1)-spacetime, but in which information operates in an extended manifold, an extended space. Samal, author of papers on bosons and neutrino oscillations and radiation and electrodynamics and gravity, concludes his 2002 "Speculations on a Unified Theory of Matter and Mind" with a proposed line of attack for understanding matter and mind in a unified way, for formulating a generalized theory of quantum information dynamics: "The line of attack involves three steps: 1) understanding emergent phenomena and complexity in inanimate systems, 2) understanding life as emergent phenomena, and 3) understanding consciousness as emergent phenomena. The attributes of consciousness can be understood only by a prudent application of both reductionism and holism. But the emergence of consciousness will be understood as an emergent phenomenon in the sense of structural organizations in the manifold of information to yield feasible structures through which we attribute meaning and understanding to the world."[369]

***

Pitkänen has followed the steps in Samal's line of attack and produced a manifold of information—embedded 4-spaces—which represent the workings of the mind. Pitkänen's embedded 4-spaces of the mind—modeled with *p*-adic mathematics—form a structured manifold, intrinsically linked to a structured manifold of embedded 4-spaces of physics' real, classical spacetime.

TGD is Pitkänen's structure through which we attribute meaning and understanding to the world, and TGD also explains our biology, our chemistry, our physics, and our connection to the universe.

Topological geometrodynamics gives us the structure of Samal's thought space and its link to the physical world.

38.
Hidden Radical Theory #1: Pitkänen's Topological Geometrodynamics

Physicist Matti Pitkänen earned his doctorate from the University of Helsinki in 1982 with his thesis on topological geometrodynamics, and he has continued developing, expanding, and promoting TGD ever since. Breaking down the name of this theory into its three components—topology + geometry + dynamics—we gain some insight into the theory's scope: TGD is a geometry-based modeling of the universe and its dynamics, with the shapes (topology) of space playing a key role in the mathematical modeling. TGD also adapts string theory to create models of the elementary particles. And it uses complexity theory's fractal geometry to model the physics of quantum theory.

Where Pitkänen takes this is pretty astounding: he takes this to a science of consciousness, and to a link between the universe and the evolution of biological systems in general and the mind in particular.

This chapter closes with a rapid-fire listing of some of Pitkänen's ideas, drawn from the thousands of pages of his work. You'll be amazed. Do not resist; you'll get the gist.

TGD: our Number One Theory of Radical New Physics. [370]

***

Normally we think of ourselves as three-dimensional beings in a three-dimensional world, and we can accept physicists' idea that time is a fourth dimension, implying that ours is a four-dimensional spacetime.

Special relativity corrects the relationships among the four dimensions, due to the speed of light being finite and constant. Physics' four-dimensional spacetime, corrected for special relativity, is called Minkowski space and is written M^4.

A problem arises when physics' notion of spacetime is further corrected, for general relativity, which understands spacetime as reshaped by matter. The problem that arises is that correcting M^4 for general relativity results in energy no longer being conserved. This is bothersome, because physicists are disinclined to give up on the conservation of energy.

Pitkänen proposes a radical notion of spacetime in order to permit energy to be conserved. Pitkänen's notion is that four-dimensional spacetime exists within a space with a higher number of dimensions, and conservation of energy is a characteristic of this higher-dimensional space. The idea is that conservation of energy takes place in all of this higher-dimensional space, and that therefore conservation of energy is also a characteristic of the M^4 factor of this higher-dimensional space, rather than a characteristic specifically of M^4 space itself.

Pitkänen goes on to show that evidence from the physicial world permits only one shape for this higher-dimensional space. It must be a space with four extra dimensions which are compact dimensions, of size ten thousand Planck lengths, or 10^{-30} meters. One way of picturing this is to stop thinking of our typically understood four dimensional world's smallest points as just points, but instead thinking of replacing these points with a compact four-dimensional space, which physicists call the complex projective space $CP2$.

Next, Pitkänen asks you to open your eyes and trust what you see. If you're looking at a person, you see a body, which contains organs, which contain cells, which contain cell organelles, which contain molecules, which contain atoms, which contain elementary particles. Trust that what you're seeing is what is there—a body that is a four-dimensional universe, ending at its outer boundary, enclosing smaller universes and enclosed by larger universes.

There is a hierarchy of universes—embedding and embedded—which are connected by elementary-particle-sized wormholes. Each universe is its own four-dimensional spacetime "sheet," with our traditional three dimensions of space plus a dimension of time. The tiny wormholes, connecting embedding and embedded sheets, exist in dimensions beyond the four spacetime dimensions, in $CP2$ space. This hierarchy of embedding and embedded four-dimensional spacetime sheets—this many-sheeted spacetime—exists within TGD's eight-dimensional embedding space.

We have not yet quite finished setting up Pitkänen's TGD universe. We need two additional characteristics of TGD, two additional principles or tools. Together, these tools allow Pitkänen to explain quantum physics and particle physics, and to explain the mind's role in physics as well as the physics of the mind.

TGD's first additional principle is how TGD understands quantum physics. In TGD, quantum physics' superposition of unrealized possible outcomes is—except for the quantum jump itself—a phenomenon of classical not quantum physics, structured as a configuration space, the space of physically allowed spacetime surfaces, which is a world of classical worlds. During the quantum jump, which is a moment of consciousness, one configuration space of physically allowed spacetime surfaces is replaced by another.

Consciousness is mathematicized into the heart of TGD physics through a special form of mathematics, called p-adic mathematics, our final tool needed to proceed with this introduction to TGD. TGD uses p-adic mathematics to model thought and to link the mind with the real physical world.

p-Adic Fractality

Pitkänen's early work created multidimensional geometric models of gravity, from which features of the other forces of physics as well as the elementary particles also emerged. Over time, Pitkänen incorporated within these efforts p-adic number fields, which are based on fields of prime numbers p that emerge from number theory and are related to fractals, complexity theory's infinite-dimensional snowflakes on snowflakes on snowflakes. Iteration at greater and greater degrees of resolution is central to fractal geometry, and this is why we'll find central applications for fractals within TGD.

Prime Numbers and p-Adic Mathematics

We've noted (Chapter 23) scientists' fascination with the effectiveness—unreasonable effectiveness, some say—of mathematics as a tool of science. Prime numbers have been one aspect of this fascination. It is not lost on physicists, for example, that the fine structure constant's 137 is a prime number.

Scientists note a whole series of physical and biological systems—from particle physics to the life cycle of species—in which prime numbers play an important role.[371] One popularly mentioned example involves a 1972 chance meeting between number theorist Hugh Montgomery and physicist Freeman Dyson, which led to introducing Montgomery's work on the spacing between prime numbers into Dyson's work on quantum dynamical systems.[372]

Mathematician Kurt Hensel first described the p-adic number system, based on prime numbers, at the beginning of the twentith century. Basing a number system on powers of prime numbers p is, at first blush, not particularly different from basing a number system on powers of ten, as our familiar decimal system does.

In fact, p-adic numbers can be written looking just like decimal numbers: 873.14, for example. If this is a p-adic number—with $p = 11$, for example—then the 7 within 873.14 is telling us how many 11s (not 10s) are in the number, the 8 is telling us how many 121s (not 100s), the 1 is telling us how many elevenths (not tenths), and the 4 is telling us how many 121sts (not how many 100ths). This is (so far) just the familiar base 11 arithmetic you learned about in junior high school.

But if you've gotten this far in the book, you've learned—as the White Queen instructed Alice—to practice believing impossible things, perhaps as many as six impossible things before breakfast. So the wonderland of p-adic mathematics is going to require imagining much more than junior high school mathematics. This is because, through the looking glass, p-adic mathematics is conceptualized backwards.

In the mathematics of real numbers, a decimal number's size is based, most importantly, on its first digit, the digit furthest to the left. For the decimal number 873.14, for example, the most important number indicating its size is the 8, which indicates that 873.14 falls within the 800s.

In p-adic mathematics, however, the furthest digit to the left is not what creates p-adic mathematics' analog to a real number's size. The furthest digit to the right does.

When we talk about a real number's size, we're probably referring to its *absolute value*. For a positive real number, such as 873.14, its absolute value is 873.14. And -873.14, a negative real number, also has 873.14 as its absolute value. In this sense, the size of either the real number 873.14 or the real number -873.14 is 873.14.

Mathematicians, who like to consider how other hypothetical number systems besides real numbers might act, consider absolute value to be just real numbers' specific implementation of a more general concept, called the norm. And the norm of p-adic numbers is very different from the norm of real numbers.

For p-adic numbers, the norm—the analog to real numbers' absolute value—has to do with the p-adic number's last digit, 4 121sts for the 11-adic number 873.14. So the p-adic number 873.14 is not very close to 870, but is very "close" to 319,498,207,695.08. Note that it's not even the value of the last digit on the right that's important in determining the similarity in size of two p-adic numbers; it's just the fact that these two 11-adic numbers are both written with two digits to the right of the decimal point.

In the p-adic numbering system, we aren't always that interested in the digits far to the left, just as in our real decimal system, we aren't always that interested in the digits far to the right. In p-adic mathematics, two numbers based on the same prime number p are the same size (more technically, have the same norm) if the two p-adic numbers continue with the same number of digits to the right.

And p-adic geometry, as you might imagine, is also very peculiar. p-Adic geometry involves only closed geometric shapes. As a result, p-adic surfaces have no boundaries. A p-adic line is always a closed loop of some shape; it has no endpoint. This is p-adic mathematics as unordered, ultrametric, non-Archimedean. There is no endpoint or boundary, so ordering can be by hierarchy of enclosure but not otherwise by size.

In p-adic geometry, either one shape is completely inside another shape, or else it's completely outside another shape, but two shapes never overlap in p-adic geometry. In p-adic geometry, what matters is inclusiveness and the level of embedding or hierarchy. Shapes enclose whole levels of shapes, and shapes share their level with other shapes but never overlap with them.

p-Adic mathematics' hierarchical geometric structure relates directly to the p-adic arithmetic that we've discussed. p-Adic numbers that contain more digits to the right geometrically incorporate p-adic numbers with fewer digits to the right. A p-adic number with many digits to the right is large, and it is portrayed geometrically as incorporating (encircling) smaller p-adic numbers, which are numbers with fewer digits to the right.

p-Adic Mathematics Finds Its Way into Science

You must be wondering why on Earth anyone would bother with p-adic mathematics. At best, it seems like a toy for mathematicians with too much time on their hands.

Pitkänen, developer of TGD, developed as an early application of p-adic mathematics a p-adic description of the Higgs mechanism which is postulated to impart mass to matter.[373] Other physicists have used p-adic numbers in various roles since the 1980s. We've mentioned Sidharth's use of p-adic numbers within the quantum black holes and fluctuational cosmology that we summarized as Radical Theory #3 of New Physics. And Castro, who looks to infinite-dimensional spacetime to explain quantum gravity, notes that ordinary real numbers are not useful for describing the infinite dimensions of fractal spacetime, but that p-adic numbers are naturally suited to fractal mathematics.[374] Others have looked at applications of p-adic mathematics to particle physics,[375] quantum cosmology,[376] the nature of spacetime,[377] p-adic gravity,[378] string theory,[379] crystals,[380] tachyons,[381] basic theoretical issues—such as the Heisenberg uncertainty principle,[382] the hidden variable interpretation of quantum reality,[383] and quantum probability theory[384]—as well as research important to quantum computing, such as neural networks[385] and parallel p-adic algorithms.[386]

Pitkänen expanded his p-adic modeling to develop TGD physics as a type of number theory. In TGD, biological systems ("biosystems") in general, and the human brain in particular, are built to mediate between regions of real spacetime, metricized through real numbers, and regions of cognitive spacetime, metricized through p-adic fields. Through the "p-adicization" of the physical world—modeled mathematically by linking a real-number-based framework and a framework based on p-adic number fields—the cognitive p-adic structures of the mind are linked with the physical structure of matter in its real spacetime universe. p-Adic spacetime is the cognitive representation of the real regions of spacetime in which matter resides.

Said another way, p-adic spacetime represents mind stuff.

Rational Numbers Link Real and p-Adic

p-Adic mathematics is a different completion of the rational numbers, a completion of rational numbers accomplished differently from how real numbers complete the rational numbers.

This sounds very mysterious, and in fact it does go fairly deep into the theory of mathematics. Rational numbers are a step more elaborate than the simple counting numbers and integers: 1, 2, 3, and their negative partners and zero. Rational numbers also include fractions—2/7, 103/1000, as well as all numbers that we write as decimals—numbers that are *ratios* of integers.

But many numbers are not rational, for example the square root of two and geometry's pi. They are not the ratios of any two integers. These irrational numbers are nevertheless real. As an aside, pi is considered so irrational that it's called transcendental, but nevertheless pi is still a real number. We've learned that we can approximate irrational numbers, even transcendental numbers, to any degree of exactness that we'd like as decimals, which are a form of rational number—the square root of two, for example, as 1.4142135624, or pi as 3.1415926536, to ten places to the right of the decimal point. But these decimal expansions—these rational numbers—are approximations of the real numbers. No matter how many digits we use, these rational (decimal) numbers will never equal the real numbers. The real numbers will always be just a little larger or a little smaller than any decimal (rational) expansion. We need the real numbers to complete the rational numbers.

Unless you're studying the theory of pure mathematics, you've probably never cared about what assumptions go into the conventions of our meaning of number. These conventions are actually mathematical models, not deep reality. One aspect of our meaning of number is the concept that the set of real numbers is a completion of the set of rational numbers—that there are more real numbers than rational numbers, and that real numbers fill in the spaces between rational numbers.

But pure mathematicians recognize that using real numbers is not the only way to fill in the spaces between rational numbers. They recognize that other approaches are possible for how we might complete the rational numbers. p-Adic mathematics is an alternative approach to completing the rational numbers, an alternative to completing the rational numbers with real numbers.

In TGD, p-adic mathematics is the mathematics of the mind, of thought, of consciousness and cognition. The physical world operates according to real mathematics. p-Adic mathematics and real mathematics are different completions of the rational numbers, but they are fused at points of common rationals, points where the completion of the rational numbers branches off toward either a real completion or a completion based on p-adic number fields. Reality and intention are fused at points of common rational numbers, fused more tightly when there are more rational numbers in common. And in TGD, our molecular biology mediates between the real physical world and the p-adic world of the mind.

The Ostrowski Theorem

p-Adic number fields have been studied extensively enough for mathematicians to demonstrate that they create a complete system of mathematics, fully internally consistent, even though p-adic mathematics has some characteristics that differ from our usual real mathematics. p-Adic mathematics is fractal. p-Adic arithmetic, geometry, and topology are fractal. And p-adic mathematics is ultrametric, non-Archimedean: distances, lengths, even the ordering of p-adic numbers is not defined in the manner that real numbers have accustomed us to. In fact, p-adic numbers are not well-ordered: for two p-adic numbers with the same norm (same number of digits to the right in their $p$inary expansions), there is no way to say which is larger. Perhaps this is why p-adic mathematics has been found particularly useful in applications involving the fine-structure constant[387]: Archimedean metricity is inappropriate in the Planck-scale fine-structure world of ultrametricity, a world at the smallest limits of what is meaningfully measurable.

p-Adic geometry has the unusual property that p-adic geometric forms are either completely nonoverlapping, or else one p-adic geometric form is completely contained within another. There are no partially overlapping shapes in p-adic geometry. A world pictured with p-adic geometry is a world in which multiple shapes share a larger space but do not overlap, and in which shapes wholly contain smaller shapes, shapes floating within shapes floating within shapes. In p-adic geometry, we find replication at multiple levels of forms encompassing forms encompassing forms. And we'll need to be prepared for unfamiliar concepts of distance and nearness, and for an unfamiliar sense of ordering.

Mathematicians and physicists note a critical aspect of p-adic mathematics. This is the Ostrowski theorem, which states that real numbers and p-adic number fields, labeled by prime numbers p, define the *only* completions of the rational numbers[388]—they are the only two number fields that contain the rational numbers as a *dense subfield*.[389]

What a temptation this creates for those attracted to mathematically derived aesthetics, or aesthetically derived mathematics: there are uniquely two systems of mathematics—the real and the p-adic—which effectively complete the rational numbers, the most basic mathematical representation of the world.

Others, besides Pitkänen, have even been tempted to consider physics that intertwines these two mathematical systems, the real and the p-adic.[390] But it is the intimate intertwining of these two basic ways to complete the rational numbers, and the broad and extensive applications—to physics, chemistry, biology, cosmology, psychology, and beyond—that are the central distinguishing features of Pitkänen's use of p-adic mathematics.

p-Adic Mathematics as the Mathematics of Thought

In some ways, it is not surprising that the strange p-adic mathematics has applications in psychology.

Consider how your mind wanders, from one thought, to another thought which feels similar but at the same time seems unrelated. Are these memories connected because they are p-adically near—because they have similar details, even if dissimilar real size or scope? And is p-adic mathematics the mathematics of not just our daydreams but our night dreams as well?

How is it that memory fragments can so often resonate for us so broadly? Do we label our mind's filing cabinets with the tails of memories—according to their p-adic norms—permitting the full memory to reawaken from just its detailed bits, permitting memory fragments to trigger past associations?

How do we store memories, reducing highly complex realities into thoughts that we will later retrieve? Is this by the p-adicization of the real sights and sounds that our senses observe?

And isn't p-adic mathematics' natural hierarchical structure ideally suited for pattern recognition processes that the human mind seems to perform so much more effectively than even the most powerful computers?

These possibilities for a p-adic mind—for p-adic modeling of thought—have been considered by a small group of mathematicians and physicists, in addition to Pitkänen. p-Adic mathematics has been used to simulate features of the thinking process,[391] to model information spaces,[392] and as a formalism for cognitive measurement.[393] Physicists have described human memory as a p-adic dynamic system[394] and have proposed an information interpretation of p-adic physics.[395]

Understanding p-adic mathematics as the mathematics of thought is a first step toward understanding TGD. The entire development of TGD physics assumes the fusion of the real physical world with the p-adic world of the mind.

Many-Sheeted Spacetime

TGD models both spacetime and matter as a hierarchy of surfaces or sheets, connected through tiny wormholes. These wormholes are the size of an elementary particle, with spatial dimensions of ten thousand Planck lengths, about 10^{-30} meters.

A larger spacetime sheet (for example, that of an atom) represents the external world from the viewpoint of a smaller spacetime sheet (for example, that of the atomic nucleus). A tiny wormhole residing near the boundary of the nucleus's spacetime sheet opens up to the spacetime sheet of the atom. In turn, a wormhole connects the world of the atom to the world of its molecule, and so on to the external world of the cell, the organ, the organism, and beyond.

Each spacetime sheet is four-dimensional and has finite dimensions and an outer surface. It is a baby universe and is encompassed by another, larger baby universe, connected via a tiny four-dimensional wormhole. Mathematically, each spacetime sheet is a four-dimensional surface within eight-dimensional embedding space.

Physics as ordinarily understood contemplates four-dimensional objects as taking up a section of infinitely extended four-dimensional spacetime. TGD contemplates something completely different. In TGD, a four-dimensional object is a finite four-dimensional universe. This four-dimensional object both encompasses and is encompassed. As the encompasser, this four-dimensional object creates an embedding space for objects that it encompasses. As the encompassed, it is an extended but finite four-dimensional universe, within embedding space created by its encompasser.

Pitkänen calls the spacetime sheet of the embedded object an associative quaternionic submanifold of octonionic embedding space. Eight spacetime dimensions are relevant for any physical object, the four spacetime dimensions in which it is embedded, plus the four spacetime dimensions that circumscribe the object as its own finite universe. The embedding is realized physically via the tiny wormholes residing in the tiny $CP2$ space that constitutes eight-dimensional embedding space's dimensions #5, 6, 7, and 8.

This is many-sheeted spacetime, an understanding of the physical world as hierarchies of enclosing and enclosed four-dimensional finite universes, thought of as two universes—the enclosed and the enclosing, eight dimensions—at a time.

Pitkänen offers this self-analytic commentary: "Many-sheeted space-time leads to the geometrization of structures and matter in terms of the macroscopic topology of the space-time surface and means that we live in the middle of the wildest science fiction."[396] Others have labeled this "Escher's dragon" when four-dimensional spacetime embedded within eight-dimensional space has been used in a somewhat different application, to model quantum gravity.[397] Like Escher's staircases that eternally ascend and descend, TGD's four-dimensional universes infinitely encompass and are infinitely encompassed.

Wildest science fiction. Escher's dragon. But also reminiscent of deep understandings of systems and how they are to be analyzed. For example, Douglas Ross developed a proprietary systems analysis technique, for the Massachusetts company SofTech, Inc. that he founded, from a conceptualization of the universe's systems similar to Pitkänen's many-sheeted spacetime. To convey the conceptual basis of this systems analysis technique, Ross produced a wonderful science-fiction-type vision, including *Plex1: Sameness and the Need for Rigor* and *Plex2: Sameness and Type*, in which astronauts of the future need to be prepared to understand the unknown, which will come in shapes, sizes, and scope that they will have to face without preconception. So their first task will be to scope out the unknown, which they are able to do only through the use of their "sameness generators," hierarchically connected detectors of the patterns embedded and repeated at multiple levels of observed reality. Ross built a business's successful systems analysis technique from this theoretical foundation of "plex" and "sameness detectors."[398]

But we're not quite done with explaining TGD's many-sheeted spacetime.

Remember: TGD uses both real numbers and p-adic numbers in modeling the topology of spacetime. The natural interpretation of the p-adic regions is as cognitive representations of real physics. p-Adic regions correspond to mind stuff, the regions of spacetime where cognitive representations reside.

So the TGD world is a world of both material and mindlike spacetime sheets—real and p-adic spacetime regions as geometric correlates of matter and mind. The p-adic world of the mind is, like the real physical world, also structured as many-sheeted spacetime, four-dimensional universes both embedded within and embedding other four-dimensional universes. And for the mindlike spacetime sheets, this embedding is naturally modeled with p-adic mathematics, the mathematics of enclosure.

TGD's many-sheeted spacetime decomposes into regions of many-sheeted real topology and regions of many-sheeted p-adic topology. Let's see where Pitkänen takes us from here.

The Quantum Jump

The configuration space of quantum physics' physically allowed spacetime surfaces is TGD's arena for quantum dynamics.

This configuration space is an infinite-dimensional world of classical worlds, an infinite-dimensional set of perceptively equivalent physically allowed spacetime surfaces that represent all quantum possibilities from which just one will be actualized during the quantum jump.

Quantum jumps occur a thousand trillion trillion trillion (10^{39}) times a second: the spacing between quantum jumps is the time that it takes light to cross dimensions five and higher, which separate the four-dimensional spacetime sheets of many-sheeted spacetime. At each quantum jump, one quantum reality is replaced by another. A quantum reality is an infinite superposition of perceptively equivalent spacetime.

The quantum jump is a cognitive process, consisting of state function reduction and state preparation by self-measurement. The process is governed by the *negentropy* (negative entropy) *maximization principle*, according to which the information content of the conscious experience is maximized.

Traditional physics' concept of entropy measures disorder, the lack of information. This is an unsatisfying measure, because often we're more interested in the presence, not the lack, of information.

Traditionally, physics gets around this by focusing on decreases in entropy measuring increases in information. But TGD goes beyond this, because p-adic mathematics permits negative entropy, which measures the presence of information. Negative entropy is also modeled in real mathematics.

Others have noted a key role for negentropy measures in neural network modeling that enhances artificial intelligence capabilities.[399] And *anti-entropy* or *counter-entropy* processes are at the heart of hypotheses linking gravity, the collapse of the quantum wave function, information, and life.[400] Negentropy has also been discussed in the context of evolutionary biology and ant societies[401] and even as a *nonequilibrium thermodynamics* model of North Korea's approach to interacting with its neighbors.[402]

The negentropy maximization principle is a basic principle of TGD and helps define the quantum reality that exists after the quantum jump. After a quantum jump, the previous superposition of perceptively equivalent spacetime sheets is replaced by a new superposition of perceptively equivalent spacetime sheets. In TGD, all of quantum physics is explained by many-sheeted spacetime, except for the quantum jump, a cognitive process and the building block of consciousness. The quantum jump is a moment of consciousness, a moment of re-creation between two quantum realities.

Approximately 10^{38} quantum jumps are experienced each tenth of a second, which TGD predicts to be the fundamental time scale of human consciousness, the duration of a psychological moment. Pitkänen's formulation of the psychological moment is reminiscent of the psychon that Eccles postulated as the basic unit of thought.

Self consists of material and mindlike spacetime sheets entangled with each other. This is reminiscent of Zohar's formulation of self as Bose-Einstein condensate. The *p*-adic (conscious or cognitive or mindlike) sheets map to the real (physical) sheets. *p*-Adic regions are cognitive representations of real physics. The self maintains its integrated identity by remaining unentangled with its environment during the sequence of quantum jumps.

Massless Extremals

The principles of TGD allow Pitkänen to create new understandings of basic questions of physics. For example, the rather odd, seemingly arbitrary distributions of the strengths of physics' forces and the sizes of physics' elementary particles emerge from the mathematics of *p*-adic numbers and the separation of TGD's spacetime sheets.

New building blocks for the physical world are conceptualized through TGD.

TGD predicts numerous instances of exotic electroweak physics and predicts the generation of coherent states of photons and gravitons. Reminiscent of advanced extraterrestrials' communication in Carl Sagan's science fiction work *Contact*, the phase of the coherent electromagnetic and gravitational radiation carries information that, in TGD, quantum antennas can transmit and receive.

Pitkänen finds biochemical evidence of *negative energy massless extremals*, which he labels "perhaps the most science fictive piece of the new physics predicted by TGD."[403] In TGD, pairs of spacetime sheets can be generated from the vacuum of space, one spacetime sheet with positive energy and the other with negative energy, so that no net energy is created or destroyed. These spacetime sheets may be *massless extremals*, which are topological light rays carrying light-like *vacuum currents* and generating coherent light. Massless extremals allow directed propagation of energy without attenuation.

Massless extremals define a fractal hierarchy starting from elementary particle length scales and extending up to cosmic length scales, acting as quantum holograms, carrying representations across hierarchical levels. They act as receiving and sending quantum antennas. Massless extremals represent topological mathematics, and they have physical representation in Bose-Einstein condensates of photons and gravitons. Massless extremals can also serve as Josephson junctions between magnetic flux tubes. And they are the building blocks of cognitive structures and are the seat of higher levels of consciousness.

Negative energy corresponds to a reversed arrow of *geometric time*. In TGD, geometric time is the real time of spacetime physics and can proceed forward or backward—that is, in a positive or negative direction. Our sequence of quantum jumps, which create our subjective experience of time, can take us to quantum superpositions of the geometric past or the geometric future. Our continuous experience of consciousness derives from the suspension of conscious experience during periods between quantum jumps. These are the periods of infinite superposed quantum possibility, with conscious experience suspended because the quantum jump—the moment of consciousness—is not proceeding. Thus we obtain a subjective experience of time from the succession of quantum jumps.

TGD and Biosystems

In TGD, biosystems are macroscopic quantum systems.

Biosystems are superconductors and superfluids and take the form of liquid crystals, which have the ability to self-organize to very complicated structures. These liquid crystals appear in various forms—biostructures—including microtubules, cell organelles, and cells. Microtubules, for example, are quantum computers, transferring information from cell membrane to cell nucleus. *Topological quantum computation* is a central TGD process.

Evolution has favored the development of biomolecules permitting these quantum self-organization processes, with biochemistry and DNA replication being important applications. DNA's code *holograms* to the larger biostructure, which scans with massless extremals.

Positive- and negative-energy massless extremals are paired and move in opposite directions along the double strands of DNA, generating *biophotons*, a type of coherent light. These biophotons then Bose-Einstein condense to create well-ordered, specific biological control mechanisms—mechanisms by which the DNA spacetime sheet is replicated to the spacetime sheets of cell biology. Thus massless extremals are associated with the electromagnetic expression of DNA's genetic information.

This is the self-organization of matter.

Pitkänen sees applications in cancer research, in pharmacology, in healing and medicine. And he sees massless extremal phenomena throughout our biology—in the magnetism of hemoglobin and in magnetic crystals in the brain, in the electroencephalogram (EEG, brain wave scans) and the nerve pulse.

The many-sheeted structure of DNA creates the miraculous features of DNA replication and cell differentiation. Genes are many-sheeted spacetime structures, faithfully coding the topology of the expression domain of the gene. A hierarchy of weakly coupled superconductors exerts quantum control and coordination. DNA and RNA transcription have their cognitive counterparts in the one-to-one correspondence between nervepulse patterns and sequences of cognitive neutrino pairs.

Biosystems are neutrino superconductors. Neutrinos are ideal candidates for *cognitive fermion pairs*, matter particles which permit entanglement which in turn is critical for memory. Neutrinos couple to the atomic nucleus via the classical weak force, not the electromagnetic force. This creates a natural insulation against electromagnetic and chemical perturbations. Thoughts correspond to the creation and destruction of cognitive neutrino pairs.

Other matter and force particles are important to consciousness. Under TGD's *quantum antenna hypothesis*, microtubules create a coherent state of biophotons and possibly also gravitons. Microtubules are senders and receivers of coherent light. The massless extremals associated with microtubules and other linear structures are for neuron groupings what radio receivers and radio stations are for us.

TGD is deeply nonreductionistic, holistic. Pitkänen notes that his applications of microtubules are different from those we saw much earlier, which had proposed "reductionist identification"[404] of microtubules with consciousness, directly identifying microtubules as the seat of consciousness. TGD's microtubules are biological structures that physically reach into the fused world of conscious and real many-sheeted spacetime to link neurobiology with intention.

In TGD, biosystems are conscious holograms, senders and receivers of remote mental interaction. *p*-Adic cognitive spacetime sheets are of astrophysical size, *magnetospheric sensory representations.* Human consciousness involves astrophysical length scales. Our personal and sensory magnetic canvases are defined by our brains, our bodies—and by the magnetic body delivered by planetary and other cosmic influences.

Brain and Thought Processes

Like other biosystems, the brain is a macroscopic quantum system.

The brain is self-organizing quantum spin glass.

The brain is a fractal associative net, structured to holographically pass information via neuronal windows. Coherent photons are emitted by mindlike spacetime sheets and propagate along axonal microtubules which serve as wave guides. Coherent light is for the brain what radio waves are for us—transmitted from a distance, for us to attend to, act on.

Mindlike spacetime sheets are massless extremals, quantum holograms. Microtubules and other linear structures act as quantum antennas and generate coherent photons.

Massless extremals define an infinite hierarchy of electromagnetic lifeforms. Extremely low frequency massless extremals are essential elements of EEGs. The EEG is a direct physical correlate of mindlike spacetime sheets associated with our extremely low frequency selves.

Radio wave and microwave massless extremals represent the mental images related to sensory experience and have sizes below brain size. Ultralow frequency massless extremals with time durations measured in the time scale of our life cycle have astrophysical sizes and make possible generalized sensory experiences about transpersonal levels of consciousness. And they make possible long-term memories.

The general philosophy behind TGD's model of EEGs and nerve pulses is the theory of quantum control and coordination based on the master/slave hierarchy of weakly coupled superconductors controlling each other by Josephson currents. The basic structure is a Josephson junction formed by two superconductors. Sensory experience is characterized by magnetic transition frequencies of various charged particles in Earth's magnetic field and in the axonal cell membrane.

This phenomenon is reminiscent of magnetoreception mechanisms by which migrating birds maintain a sense of direction through interactions between the Earth's magnetic field and magnetic crystals in the birds' bodies.

Stone age man received the commands and advice of a collective consciousness as auditory and visual hallucinations via regions of the right brain hemisphere, from which these commands were communicated to the left hemisphere. By contrast, modern man receives these communications as thoughts (internal speech) in left brain semitrance and as emotions in right brain semitrance.

Sensory representations are realizations outside the brain. Information cannot be located within the brain in any specific location.

The brain is a Boolean net, a network of on/off switches interconnected with logic gates. Nerve pulse patterns are coded to Boolean statements. TGD models the role in the mind's Boolean cognition of the cognitive neutrino pair—a neutrino and an antineutrino, with vanishing total energy, residing at different spacetime sheets, but quantum entangled. This entanglement makes possible cognition, implemented by cognitive neutrino pairs.

Negative-energy massless extremals are the topological correlate of binding energy, and Pitkänen sees the holistic right hemisphere of our brain generating negative-energy massless extremals which bind its magnetic bodies into a single whole, in contrast to the analytic left hemisphere's decay into large numbers of subselves.

Memories

Gravitonic massless extremals are responsible for memories.

There are no memory storages in the brain. Only memory representations abstracting the essential aspects of experience are needed. The identification of memories with spacetime sheets makes it unnecessary to store information about the geometric past to the geometric now.

Synapses code cognitive representations and learned associations, not information itself about events in the past. Mental images of the geometric past are entangled with mental images in the brain.

Long-term memories are geometric memories. Long-term memory is coded in the classical electromagnetic field and in coherent light generated by massless extremals in a hologram-like manner.

Water clusters and macromolecules with sizes in the range of cell membrane thickness are good candidates for generating gravitonic massless extremals responsible for all geometric memories.

The narrative character of long-term memories is a multitime experience, experiences bound together via quantum entanglement.

A single memory is a single spacetime sheet, without quantum jump. A mirror model allows the self of the geometric now to see the self of the geometric past. The quantum jump between different classical gravitational configurations involves the emission of gravitational massless extremals, and the reflection of the gravitational massless extremals from a curved vacuum spacetime sheet realizes the miror mechanism.

Information content is the change in information associated with initial and final quantum histories, the positional information gain associated with a moment of consciousness.

Sensory representations are realized on magnetic sensory canvas having size much larger than the physical body. These are magnetospheric sensory representations.

The cosmology of consciousness has fractal-like structure, with subcosmologies which know nothing about each other's existence except as a result of quantum jumps involving entanglement with larger spacetime sheets.

Anticipation of the future results from massless extremals p-adically linked to intentions, plans, and expectations. This allows simulations of what might happen in the future (and would have happened in the past) without quantum jump.

Reaching Out to the Universe

Pitkänen looks at flux tubes of Earth's magnetic field and he looks at the Northern lights. He finds an explanation for the universe's dark matter, and he predicts higher levels of consciousness.

The universe is an algebraic hologram. Symmetries are replicated as fractal scalings. Biomatter mimics the Earth's magnetic field, communicating the replication of topology through wormholes across many-sheeted spacetime.

The structure of many-sheeted spacetime represents the structure of the cosmology of consciousness. Our individual development and the development of our species recapitulate developments throughout many-sheeted spacetime. Both past and future civilizations participate in each quantum jump.

In the TGD framework, the highest levels of our personal self hierarchy correspond to sizes much larger than the physical body. The astrophysical size of our cognitive spacetime sheets implies that human consciousness involves astrophysical length scales.

Pitkänen looks at neurophysiological evidence for the resonance in the brain of geomagnetic fields and extremely low frequency electromagnetic fields. He finds observations of special effects in the EEG frequency range and is inspired to hypothesize that our selves correspond to topological field quanta that have size of the scale of Earth. Pitkänen concludes: "This leads to a rather radical modification of the brain-centered views about consciousness."[405]

Living matter is known to be particularly susceptible to extremely low frequency electromagnetic fields, which, by the Heisenberg uncertainty principle, can be shown to correspond to large topological field quanta. For example, photons in the ten hertz frequency range generated by Josephson currents correspond to topological field quanta of the size of Earth. This suggests biological development that is attuned to planetary effects. Also, the cyclotron frequencies of various charged particles in Earth's magnetic field correspond to time scales important to various neurological processes.

Microtubules and other biological structures act as quantum antennas—waveguides for photons of coherent light. Quantum antennas are neural windows, using the same topological quantum computing processes as the holographic brain.

We are much more than our neurons. TGD is not a theory blinded by standard mechanistic and reductionistic prejudices about the brain.

A Jackpot of Private Thinking

Matti Pitkänen's far-reaching and robust website includes some personal reflection, ominously labeled his "Doomsday diaries." TGD is far from universally accepted by physicists. A 2000 entry in Pitkänen's Doomsday diaries refers to a physicist colleague of Pitkänen who has recently referred to Pitkänen in a Finnish journal: he has labeled Pitkänen a "private thinker," saying that the physics department does not in any way support Pitkänen's work. Pitkänen comments: "Let us add that 'private thinker' has quite special meaning in the language of Finnish scientific community: the faithful translation would be something like 'crackpot.'"[406]

Pitkänen is a forceful advocate of new physics that is dissident physics, not Big Science. And he wonders if today's generally accepted academic theories of physics take insufficient notice of both longstanding and new observations that contradict today's accepted tenets. About one recent new physics observation Pitkänen writes: "Certainly, at the times of Einstein empirical discovery of this caliber would have been a major scientific event."[407]

Para-Applications

As you read about the components of TGD, chances are your mind wandered toward additional applications in science and psychology, and toward applications in parapsychology and the occult.

Pitkänen's work is peppered with para-applications, if I can coin a phrase. These probably crossed your mind as you read the summary above of elements of TGD science.

Pitkänen, as he writes about TGD, makes connections with ESP, precognition, dreams and hallucinations, teleportation, religion, how the musical scale is coded into axonal anatomy, communication between the living and the dead, reincarnation, karma.

He looks to the astrophysical phenomenon of auroras for what they tells us about atmospheric superconductivity and the dynamics of Earth's magnetosphere.

He contemplates crop circles as a dramatic example of many-sheeted spacetime, reflecting extraterrestrial communication from parallel spacetime sheets. Or perhaps crop circles are the work of intraterrestrial aliens—ITs—who exercise control over the magnetic fields associated with earth's inner core.

Pitkänen also discusses aging, illness, and near-death experiences. Healing is tachyonic radiation, a superluminal phenomenon creating time reversal of biological processes. Homeopathic remedies and acupuncture involve electromagnetic frequency signatures of chemicals that can be understood if homeostasis is based on many-sheeted ionic flow equilibrium. Schizophrenics receive as hallucinations commands and advice from collective levels of consciousness. Aging is the price for having self. Can some electromagnetic selves survive physical death? Can we generate artificial life by generating electric fields?

Pitkänen discusses emotions, altered states of consciousness, quantum ethics, good and evil. Good deeds are those that support evolution, p-adic evolution—p-adicization with more detail and with higher prime numbers p.

Evolution is the evolution of mathematical structures, which become more and more self-conscious quantum jump by quantum jump. Short real lengths correlate with large p-adic scales. This implies physics' progress in mastering shorter and shorter real lengths accompanies an evolution in the human species' consciousness.

Living systems exhibit a strong presence of electromagnetic fields, which play important roles in living matter. Magnetic fields are generated by biological features such as the eyes and the pineal gland. The pineal gland is located behind the middle of our forehead and seems vestigially related to structures in reptiles that are more clearly eyelike. Mystics consider the pineal gland to be the site of our "third eye" of extrasensory perception. The frequencies of the eyes' and the pineal gland's magnetic fields correspond to human time scales. Pitkänen relates these magnetic fields to near-death experiences, suggesting that these experiences could survive after death.

Two modes of consciousness can be identified: ordinary and whole-body consciousness. These modes could naturally explain rational/analytic and emotional/holistic modes of mind. These two modes could make it possible to understand various dichotomies like left/right brain, analytic/emotional, rational/religious, Western/Eastern. Two modes of conscious existence could even provide quantum-level explanation of sex.

There is a spectroscopy of consciousness, with potential applications in physics, biology, psychology, and the development of intelligent systems. The spectroscopy of consciousness is for brain science what atomic spectroscopy has been for physics and chemistry. Stages in the spectroscopy of consciousness include high-precision neurofeedback, control of robots or artificial body parts using mere thought, direct sensory cognitive and emotional communication between brains using amplification of EEG patterns, and technologies helping to achieve various alternative states of consciousness such as meditative and transpersonal states, and states of group consciousness.

Pitkänen sees implications of TGD for our collective consciousness, for collective memory, collective learning, and for our evolution as a species. TGD predicts standing EEG waves which would correspond to transpersonal states of consciousness approaching cosmic consciousness. TGD envisions communicating through negative energy massless extremals with both the past and the future. Real-time chats can take place across a separation of billions of light years.

Concepts are living creatures. During sleep, collective consciousness begins to dominate and brains form a highly synchronous whole. Semitrance is a basic mechanism of communication between collective consciousness and individual. Consciousness survives physical death.

***

Look back at this book's Table of Contents for Part Three, New Physics. TGD has it all. Quantum gravity through spin glass analogy. Quantum jumps as moments of consciousness. A widely entangled universe of multi-sheeted spacetime in which wormholes connect the physical and conscious worlds. A world of Bose-Einstein condensates and Josephson junctions and neutrinos. A world of fractal geometry in which p-adic mathematics is core to consciousness's role in science. A world of new physics and the mind.

Epilogue
Type 1, 2, 3 Worlds

Astrobiology and exobiology and bioastronomy. Search for extrasolar planets. Search for extraterrestrial intelligence.

These emerging branches of science and near-science evoke deeply felt desires in many and deep scorn and ridicule in others.

This is progress, for those who want to believe. Today the possibility of extraterrestrial intelligent life is not the sole province of weekly supermarket tabloids.

What is today's scientific view of how probable extraterrestrial intelligence is?

What does this mean for humanity? And what role can mind physics play?

***

The Drake Equation

Astronomer Frank Drake helped organize a 1961 conference at which scientists discussed the possibility of the existence—within our galaxy—of civilizations that are trying to communicate with us. Drake was very enthusiastic about having the opportunity to discuss this with a group of fellow scientists, particularly the exceptionally qualified group that had been assembled.[408]

Drake considered how he might organize this discussion so that the group could stick to an orderly agenda and meaningfully draw conclusions. It struck Drake that, if he could break the question down into meaningful components, the group could apply their scientific knowledge to these components one by one in order to draw a rational conclusion about the full question: how many civilizations will we detect in the Milky Way?

If N is the answer to this question, then Drake suggested that the assembled group of scientists could determine N by multiplying out these seven components:

$$N = R f_p \, n_e f_l f_i f_c \, L$$

We'll discuss one by one what these scientists had to say about these components of what is now called the *Drake equation*. But first, let's remind ourselves that zero multiplied by any number is zero. So if any of the seven components of the Drake equation turns out to be zero, the scientists would have to conclude that it will be impossible to detect an extraterrestrial civilization in our galaxy.

The first component, R, represents the rate of star formation, how many stars are born each year within the Milky Way. Why this is important will become clear later on in the formula. The consensus view among the scientists assembled in 1961 was that $R = 1$. On average, one star is born each year in the Milky Way.

Next, f_p, how many of these stars have planets revolving about them. Here, there was some disagreement, so ultimately a range was settled on: somewhere between one fifth and one half of the stars have planets, are part of a solar system with planets revolving about a sun.

Moving on, n_e tells us for every solar system how many planets we'd expect to be hospitable to life. The group first agreed that for us, our solar system, this answer was three—Venus, Earth, and Mars. These planets are considered hospitable to life. Earth, of course, is particularly hospitable, but the group also agreed that Venus and Mars offered some chance of permitting indigenous lifeforms. No one felt that our other six planets would permit life. The group also felt that our solar system is probably typical of other solar systems in the galaxy, and they settled again on a range, this time that a Milky Way solar system could expect somewhere between one and five planets hospitable to life.

Next, f_l, of the planets that are hospitable to life, on how many would life actually evolve? Here, some scientists put forth the argument that if life could evolve, it would evolve. In other words, $f_l = 1$. Life would evolve on any planet hospitable to life. This became the consensus position, as high as it can get: $f_l = 1$.

Three more factors to go. The next, f_i, represents the fraction of planets with life on which intelligent life would evolve. Drake reports that the conversation here turned to dolphins. This was, after all, the era of the popular television series *Flipper*. Dolphins, it was reported, can learn complex tasks, including rescuing pilots whose planes are downed over oceans, scouting out submarines, and delivering explosives for detonation in enemy harbors. Consensus emerged that these tasks have to be considered examples of intelligent behavior, and that intelligence has great survival value for species no matter where they are located. Again, consensus emerged on the highest possible value for this factor: $f_i = 1$. Where there's life, there will be intelligent life.

The next-to-last factor, f_c, represents, for planets that have evolved intelligent life, on what fraction of these planets would there be a desire and ability for interstellar communication. Here, we must remember that communication can be accidental, not necessarily deliberate. As we've seen in many science fiction plots, such as Carl Sagan's *Contact*, Earth may first be noticed extraterrestrially because of our television signals, which should be fairly easily received if they contact any non-Earth receptor on their journey at the speed of light away from planet Earth. (Sagan was a participant at the 1961 conference where the Drake equation was discussed.) The ultimate consensus for f_c was again a range, this time between one tenth and one fifth: between ten and twenty percent of intelligent civilizations will achieve interstellar communication.

Finally, L, the longevity of a civilization capable of interstellar communication. L closes the loop on the Drake equation, which began with an estimate of R stars *per year*. Step by step, we've moved all the way from stars to planets on which intelligent life produces interstellar communication. But we're still measuring this on an annual basis, *per year*. We need to introduce longevity—multiply our previous factors by L—in order to get our final Drake equation N, the number of Milky Way planets that we can expect today have intelligent life producing interstellar comunication.

At the 1961 conference, this generated a lot of discussion: once a civilization achieved modern technology, how long would it survive? Would social advancement accompany scientific advancement, resulting in millennia of peace and progress? Or would social progress lag far behind—will the society blow itself up? And what about natural calamities, planet-based or from outer space? Will earthquakes or crashing asteroids or comets destroy the civilization before technology can control these catastrophes?

The consensus view of L settled on a dual solution: advanced technological civilizations would either decline rapidly, or flourish for eons. L could be either less than a thousand years or perhaps hundreds of millions of years.

At this point, Drake noted for his fellow scientists: "We've reached a conclusion. Our best estimate is that there are somewhere between one thousand and one hundred million advanced extraterrestrial civilizations in the Milky Way."[409]

Drake reached this conclusion—very simply, that $N = L$—by looking back at the consensus view of the various components of the right side of the Drake equation. Except for L, the factors are either one or close to one. And of the factors that are close to one, we've got n_e that could be as high as five, offsetting f_p and f_c that are in the one-fifth-to-one-half and one-tenth-to-one-fifth ranges. So as a rough approximation, recognizing that the estimate for L too is very rough, we can summarize the whole session as concluding neatly that $N = L$. The number of planets in the Milky Way with communicating civilizations is equal to the number (in years) representing the expected longevity of an advanced technological civilization. Still a very wide range—one thousand to one hundred million communicating extraterrestrial civilizations in the Milky Way, our galaxy, our neighborhood—but likely to be a fairly sizable number, and very notably not zero.

Drake notes that, while consensus as to the value of the various components has changed over the years, the overall solution for N has remained fairly consistent. Which leads us to ask . . .

Where Is Everybody?

British physicist Stephen Webb has explored the question of the existence of intelligent extraterrestrial life in his 2002 book *If the Universe Is Teeming with Aliens . . . Where Is Everybody?: Fifty Solutions to the Fermi Paradox and the Problem of Extraterrestrial Life.*

This "Fermi paradox" derives from a question—where is everybody?—that physicist Enrico Fermi posed while having lunch with some Los Alamos colleagues in 1950. Fermi, who won the 1938 Nobel Prize for physics for his work on the atomic nucleus, is also known for his contributions to the Manhattan Project and the development of the atomic bomb.

As Webb tells the story, Fermi's "where is everybody?" was immediately understood by his colleagues to be a question about extraterrestrials. The question has since been anointed the Fermi Paradox, and has fascinated Webb and many others for years.

In spite of the ridicule Webb might face in writing a book on this subject, he takes the subject seriously and draws upon centuries of philosophy, theory, research, evidence—and science fiction. Webb's work represents one example of the emerging seriousness with which scientists now approach the search for extraterrestrial intelligence.

In *Where Is Everybody?*, Webb discusses forty-nine possible solutions to where intelligent extraterrestrials are, which he organizes into three different categories: they are here or have been here in the past, they exist but have not yet communicated, and they do not exist. Webb's fiftieth solution is his own conclusion as to the solution to the paradox.

Webb fully covers the territory of possible solutions to where the extraterrestrials are. For example, within Webb's first category—they are here or have been here in the past, which Webb considers the most popular set of views among the general public—are discussions of how extraterrestrial aliens may be present in our midst today, or may have been present at an earlier time, on Earth or elsewhere in the solar system. Or perhaps we Earthlings are zoo creatures for an advanced civilization. Or maybe we are aliens: life is transported throughout the universe by panspermia, "seeds everywhere."

Webb presents variations on all of these. He takes them seriously, but in doing so finds contradictions, questionable motivations imputed to the extraterrestrials, questionable evidence, or simply reflections of our own wishes or fears. Ultimately Webb discounts all of the solutions proposed for his first category, or at the very least remains skeptical of those solutions he feels cannot currently be disproved.

Webb's analysis of his second category—they exist but have not yet communicated—is equally exhaustive. While the vast distances of space do plausibly explain why we may never have been visited by extraterrestrials, Webb is skeptical that the vastness of these distances would also explain why there would have been no communication at all. He does feel confident that we have the technology to detect an extraterrestrial communication, and does feel that the lack of successful detection of an extraterrestrial communication in spite of the search efforts undertaken—having searched billions of channels across wide ranges of the electromagnetic spectrum—rules out many proposed solutions.

It is in Webb's third category—they do not exist—that his own view lies. Webb summarizes a number of solutions in this vein: extraterrestrials do not exist because life is rare, or because life-friendly planets or technology or language and other tools of intelligence are rare. A lot of these solutions derive from picking off one by one the components of the Drake equation, explaining why a component is zero or at best exceedingly small.

Ultimately, Webb puts forth—as his fiftieth solution, "The Fermi Paradox Resolved . . ."—his own view of why communicating extraterrestrial life does not exist. Webb explains this solution as the *Sieve of Fermi*, in which he starts with an estimated trillion planets in the Milky Way, then sifts these down until he arrives at just one—Earth. A trillion planets in the Milky Way is something of a consensus number—

one hundred billion or more stars in the Milky Way, each with an average of ten or somewhat fewer planets.

Webb's sieve is conceptually similar to the Drake equation, but instead of deriving N using a time-dependent approach—based on the annual rate of star development, and the number of years for a technological civilization's survival—Webb simply starts with today's count of stars and planets and cuts back from there.

Webb estimates that elementary cellular life will evolve on five hundred thousand of our galaxy's planets—0.00005% of the Milky Way's trillion planets. He reaches this estimate by detailed analysis of components of his "sieve," examining the extent of the Milky Way's galactic habitable zone (away from environmental violence, especially of the galaxy's central regions), stars' lifetimes and stars' energy and light emission, planets' location in a zone within their solar system that is continuously habitable for the required billions of years—orbitally stable, with both water and solid ground, with acceptable heat ranges—and then actually developing elementary cellular life. So, in Webb's view, 99.99995% of the Milky Way's trillion planets are not friendly enough to proceed as far as simple cellular life. On only five hundred thousand planets—which are located in a continuously habitable zone orbiting a sun-like star which is located within the galactic habitable zone—does even simple life develop, having avoided any of various planetary disasters.

Webb proceeds with his sieve, analyzing the likelihood of these simple lifeforms developing into more complex lifeforms, then these more complex forms advancing into modern cellular structure. Webb's analysis here leads to his estimate of a 2% success rate—complex lifeforms will evolve on only ten thousand of the Milky Way's planets. And here Webb makes a final leap, that this number of planets is not enough for us to expect any besides Earth to go through all of the additional unlikely steps of high-level intelligence, abstract reasoning, and development of language and technology.

There are a lot of leaps here, and not everyone will agree with Webb. However, his classification and discussion of the forty-nine possible solutions is analysis that is important for pushing forward scientific discussion of possibilities for extraterrestrial intelligence, even for those who don't agree with how he settles the argument at solution 50.

The Zoo Hypothesis

One who does not fully agree with Webb's conclusions is astrophysicist Peter Ulmschneider of the University of Heidelberg. Ulmschneider's 2003 book, *Intelligent Life in the Universe: From Common Origins to the Future of Humanity*, summarizes in extraordinary detail decades of scientific study of galaxy, star, and planet formation; the origin of the chemical elements; the origin and evolution of life on earth; and the search for extrasolar planets and for extraterrestrial intelligent life.

This search is a long and slow process. As of 2002, only ninety extrasolar planets have been discovered, although Ulmschneider is hopeful that new search strategies will accelerate these discoveries, the first of which was as recent as 1995. Ulmschneider discusses in detail many of the subjects also discussed by Webb, including continuously habitable zones, organic chemistry and the nature of life, biological and cultural evolution, interstellar travel, the survivability of civilizations, and the Drake equation.

Ulmschneider reviews five major scientific studies, one of which is his own, that attempt to determine the Drake equation's N—the number of communicating extraterrestrial civilizations in the Milky Way. Ulmschneider's own estimate is four thousand. Another of the five studies reported by Ulmschneider is less optimistic than Ulmschneider and more in line with Webb: this other study suggests virtually no chance of intelligent extraterrestrial societies in the Milky Way. The remaining three studies summarized by Ulmschneider produce two estimates of a million and one estimate of two million communicating extraterrestrial civilizations in the Milky Way.

Ulmschneider recommends several refinements to these analyses, suggesting that his four thousand estimate isn't far from a near-consensus scientific view. Ulmschneider states baldly that "the great majority of scientists accept that there must be extraterrestrial intelligent beings in our galaxy, and certainly in the universe."[410]

Ulmschneider takes this a bit further, by noting that his four thousand planets with current intelligent extraterrestrial life imply an average distance from Earth of 1,700 light years. Based on this average distance, he discusses implications for search strategies and for the possibilities of near-term contact.

Ulmschneider reports that the first phase of the Million-channel ExtraTerrestrial Assay (META I), which ran from 1985 to 1995, produced thirty-four "alerts," signals with characteristics matching the project's screen for intelligent extraterrestrial signals, but for which the data collected did not permit definite conclusions. META II, the second phase of this search, produced nineteen similar alerts. And the search is continuing with BETA (the Billion-channel ExtraTerrestrial Assay) and SERENDIP (Search for Extraterrestrial Radio Emissions from Nearby Developed Intelligent Populations), using observatories located throughout the world.

Ulmschneider's concluding thoughts on where this will lead involve the zoo hypothesis, a theory developed earlier and one that Webb discusses (and ultimately dismisses) in the category of theories that assume extraterrestrials are here today. The zoo hypothesis asks us to consider the motivation and perspective of an advanced extraterrestrial civilization, and concludes that the inhabitants of such a civilization may well decide not to interfere with our civilization. Fans of *Star Trek* will recognize this as the prime directive. It is motivated by allowing an emerging civilization's unique development to proceed without influence, and also by fear of an uncontrolled response by a less developed civilization if it detects extraterrestrials before it is capable of handling such contact. An intelligent extraterrestrial civilization is bound to have a highly developed sense of responsibility: if it survived long enough to develop advanced technology, it must have also developed the discipline and culture to take into account others' needs and capabilities. Such an intelligent extraterrestrial civilization is likely, according to the zoo hypothesis, to hide its existence from us, satisfied to observe us—like animals in a zoo—until we are more advanced.

$N = L$

Recall the early conclusion about the Drake equation, from the 1961 conference: the number N of communicating extraterrestrial civilizations in the Milky Way is roughly equal to the longevity in years L that a technological civilization is likely to survive. The 1961 conferees established a dual consensus for L (and thus for N): roughly, either one thousand or one hundred million communicating extraterrestrial civilizations in the Milky Way.

This leads to the question of how we—the creatures of Earth—make one hundred million a more likely value for our L than one thousand. Addressing this question is one of Ulmschneider's innovations: he asks us to think carefully about the future of the development of humankind and our length of survival L, in order to gain insight into the nature of extraterrestrial civilizations and their likely number N within the Milky Way.[411]

How do we avoid the various dangers facing humankind? Ulmschneider discusses bacterial or viral infection, extreme volcanism, irreversible glaciation or a runaway greenhouse effect, comet or asteroid impact, supernova explosions or gamma ray bursts, irreversible environmental damage, uncontrollable inventions, war, terrorism, irrationality.

Ulmschneider also cites as sources of hope three areas of scientific development—information technology, the exploration of outer space, and mastery of the biological world. And as a more distant, but nevertheless foreseeable possible development, Ulmschneider discusses the connected society, a next stage of evolutionary development, in which a super-consciousness connects humankind.

Type 1, 2, and 3 Worlds

Russian astrophysicist Nikolai Kardashev has developed a classification scheme indicating how highly developed a civilization's technology is. Here, the reference to civilization means a whole planet, a whole world, a planetary society. So before we even start, it seems pretty apparent that Earth will likely be at the least developed classification, having not yet achieved a unified planetary civilization.

Kardashev proposes three types of civilizations, based on the extent of energy consumption and access to energy resources. A Type 1 civilization (also called a K1 civilization, after Kardashev) uses only the energy available to it from natural sources on its planet, such as oil and water. A Type 2 (K2) civilization is more advanced, using all of the energy radiated from its sun. And the most advanced category, a Type 3 (K3) civilization, is capable of tapping the energy of its entire galaxy.[412]

As surmised, we on Earth are a Type 1 civilization, with some very early elements of Type 2 by virtue of our beginning to make use of solar energy that has radiated to the Earth's surface. But a true Type 2 civilization would deploy energy resources far in excess of our current generation of solar energy. An example of the technological advancement of a true Type 2 civilization is the hypothetical Dyson sphere, which a Type 2 civilization would construct from parts of planets in its solar system, reconfiguring this material as a sphere enclosing its sun. This would utilize a billion times more of the sun's energy than the amount of the sun's energy that Earth intercepts.[413]

Presumably, an advanced Type 2 civilization could, as it progresses towards Type 3, travel around its galaxy constructing Dyson spheres around stars it visits. Or maybe Type 3 civilizations, with enough energy to travel through cosmic wormholes, would mine energy from the quantum vacuum or from black holes. Our astronomical observations suggest that no Type 3 civilizations exist within the Milky Way, since chances are we'd notice a civilization with energy capability of the magnitude of the entire galaxy.

The Planetarium Hypothesis

Kardashev types are classified based on access to energy.

But new physics has established energy's equivalence to information. So let's consider information-based Type 1, 2, and 3 worlds.

Webb discusses a fascinating application of Type 1, 2, and 3 civilizations' deployment of the information equivalent of energy. This is one of Webb's possible solutions to the Fermi paradox—the planetarium hypothesis.[414]

The planetarium hypothesis is in Webb's category of solutions in which it is proposed that extraterrestrial civilizations are in fact present today on Earth. The planetarium hypothesis is a variation on the zoo scenario: instead of Earthlings unknowingly residing in superior beings' zoo, we reside in their planetarium, with an artificial painted sky whose construction is so sophisticated that we don't know it's fake. Webb asks: how much information would these superior E.T.s have to use in order to construct a convincing planetarium?

Webb gives three answers—for Types 1, 2, and 3 civilizations—based on information equivalents to their respective energy levels. This requires use of the *Beckenstein bound,* which calculates the maximum amount of information that a physical system can code. This amount depends on the system's size and mass, producing as the Beckenstein bound a few million trillion trillion trillion (2.5×10^{43} to be exact) bits of information for each meter of radius and kilogram of mass. An energy/mass equivalence can be derived from the equation $E = mc^2$. And then Webb equates extraterrestrial civilizations' energy levels to various size planetariums that they could simulate around us, based on the amount of information each planetarium would have to encode. The conclusions:

A Type 1 civilization could build a convincing planetarium that would extend over the range of one of history's early civilizations, the Sumerian empire. Webb quantifies this as: ten thousand square kilometers of the Earth's surface, with a depth of two hundred meters down and a height of eight hundred meters into the skies. Include a few painted sky objects, and those Sumerians would never figure it out.

A Type 2 civilization could construct a convincing planetarium that, Webb says, would cover the voyages of Christopher Columbus, but not Captain Cook. Partially but not fully encircle the globe.

And a Type 3 civilization would create a convincing planetarium that would extend from Earth in every direction to a radius 2½ times the distance from the sun to Pluto, the solar system's most remote planet. Even today, we would not know if we lived in a planetarium that a Type 3 civilization had built around us to simulate our world.

New Physics and the Mind

I take this as a segue to something very grand about science.

What if the physics of information, and the physics of the mind, are the most basic applications of the tools of energy, mass, space, time, forces, and matter? Is physics of the mind the most grand scheme of twenty-first-century physics? Will our access to information, and to our minds, take us to Type 3? Will increased information about the cosmos and the submicroscopic—the universe large and small—also increase our access to the universe's energies? What do we need to do in order to accelerate our development of new physics of the mind?

Is it conceivable that today we are in contact with intelligent extraterrestrial civilizations throughout the universe? Are our biosystems built to sense communications from elsewhere? From extra dimensions? Through a manyfold universe? Via far-distant entangled particles? By the mind's p-adic processes in many-sheeted spacetime?

Are all of your thoughts original, or is that Platonia—or E. T.—calling?

An Optimistic Conclusion

There are several factors pointing to the possibility of some major advancements in physics being achieved in the early twenty-first century.

Confirmation of the existence of extra dimensions would be an extraordinary event of science, especially if these extra dimensions are large.

Because of these extra dimensions, and perhaps even without them, a theory of everything, as well as other major advances in our understanding of physics, may be in reach with the next generation of particle colliders and the next generation of theoretical physics.

But in my view the single event of physics that is the greatest source of optimism is the possibility of a theory of everything that, like topological geometrodynamics, includes a fully integrated theory of the mind.

Such a theory is the greatest source of optimism because of where it could take us, the human race.

By taking a theory of everything to accessible energy levels, we move entirely our sense of the technological advances that still need to be accomplished before we can make great new progress.

And by bringing in the exciting topics of new physics, as well as consciousness and the mind, we also open up whole new arenas for scientific advancement.

And, dare I say? Will we get extraterrestrial assistance in meeting our technological needs?

The dystopian vision here is that everything that I'm hinting at is true, but that we as a species can't handle it.

But suppose we embrace the utopian vision.

Or maybe the facts are all true, and the future—dystopian and utopian—is a quantum future.

Afterword
by Matti Pitkänen

Finnish physicist Matti Pitkänen, the developer of topological geometrodynamics (TGD), discusses TGD and its implications.

***

New Physics and the Mind

Mr. Paster's book finds TGD at the intersection of new observations of physics and the physics of consciousness and the mind. I am happy that the author has seen the trouble of going through all this material and managed to form an overall view about it and to express it in layman's language. I find that the series of radical new physics theories reflects also the evolution of my own ideas with all its wild twists and turns and returns to reality.

My own route to TGD as I now understand it began about twenty-five years ago from an idea which led to an attempt to unify the fundamental interactions. My own "great experience" roughly seven years later led to the realization that neither TGD nor any other physical theory I knew at that time could say anything interesting either about this experience or consciousness in general. Also the grave diffculties met in the attempt to develop quantum TGD stimulated questions about the nature of time and consciousness.

Around 1995 I began serious whole-daily work in an attempt to develop a quantum theory of consciousness, which meant that I began to write a book about consciousness. After almost a decade of writing and rewriting, I have the feeling that TGD-inspired theory of consciousness and of quantum biology exists in a reasonably well-defined sense and not only explains but also predicts.

TGD in a Nutshell

If I should express in a nutshell what is new in TGD, my statement would be roughly as follows. TGD leads at the level of physics to a new view about spacetime, time, energy, quantum, and information. TGD also leads to a physics of consciousness and cognition with an ontology of physical and conscious existence resolving the paradoxes of quantum physics. At the level of mathematics TGD leads to an infinite-dimensional generalization of the notion of geometry, to the introduction of p-adic number fields to the arsenal of theoretical physicists, and to a generalization of the notion of number. In the following I try to give a bird's eye view about the basic structure of TGD.

TGD as a Solution to the Energy Problem of General Relativity

The basic idea of TGD was stimulated by the attempt to resolve the energy problem of general relativity. Energy is not a well-defined notion in general relativity, whereas the empty spacetime of special relativity with its nice symmetries explains the basic conservation laws of energy and momentum. Gravitation makes spacetime curved and this means the loss of these conservation laws.

The fundamental creative flash was that perhaps spacetime is a four-dimensional surface in some higher-dimensional space having the same symmetries as the empty spacetime of special relativity. If so, the fundamental symmetries like translation symmetry would not be symmetries of spacetime but of this higher-dimensional space, and the notion of energy would be saved!

Having listened to basic courses in particle physics and learned the power of symmetries in quantum physics, I realized that this hypothesis would have an enormous predictive power. Once this higher-dimensional space is fixed, the elementary particle quantum number spectrum is fixed too, and it soon became clear that there is just a single candidate. Take empty four-dimensional Minkowski space M^4 and replace every point in this space with the so-called complex projective space $CP2$ so as to obtain eight-dimensional space $M^4 \times CP2$. This theory can be regarded as a generalization of superstring models which became in fashion shortly after the publication of my thesis: strings are replaced with three-dimensional surfaces in the TGD framework.

What is new is that the four-dimensional spacetime surface is looked at from the outside, from the perspective of eight-dimensional space, just as we see a two-dimensional membrane from a three-dimensional perspective. The shape of the spacetime surface as seen by this fictive habitant of eight-dimensional space gives the new degrees of freedom allowing the unification of the fundamental interactions in terms of geometry, thus realizing the dream of Einstein.

Many-Sheeted Spacetime

The attempt to formulate and understand the submanifold physics led gradually to a generalization of the spacetime concept. The notion of many-sheeted spacetime emerged gradually and I realized that every object, be it quark or galaxy, is represented by a spacetime sheet, which is essentially a universe which ends at the outer boundary of the object, and that these spacetime sheets form a hierarchy with sheets glued together by elementary-particle-sized wormhole contacts and by what I call join-along-boundaries bonds (much like threads connecting two objects). We see the many-sheeted spacetime in its full topological complexity but have not realized that we do so! It is, however, quite a challenge to express what is obvious using the language of physical measurements.

A good example about intellectual inertia making this challenge so difficult is that, although the notions of dark matter and dark energy are very familiar to me as a physicist, I realized only recently that they are direct signatures of many-sheeted spacetime—mostly ordinary matter but at larger spacetime sheets and visible only in special circumstance. This dark matter and energy turn out to be fundamental also for the understanding of living matter.

Entering the World of Classical Worlds

Already the foregoing must sound rather complicated even in the ears of average colleague but this is not yet all. In order to formulate quantum theory an even more abstract notion had to be introduced. Here I could however follow the footsteps of John Archibald Wheeler who introduced the notion of Super Space as the space of three-dimensional geometries.

Dynamical spacetime allows realization of the dream about geometrization of basic interactions. Being dynamical itself it cannot however serve as the fixed arena of quantum dynamics. Rather, spacetime itself becomes the dynamical object, and the space of physically allowed spacetime surfaces, the "world of classical worlds," becomes the arena of quantum dynamics.

The generalization of Schrödinger amplitudes in three-dimensional space to a spinor field in the world of classical worlds is quite a challenge and initiated the still-continuing program of generalizing the notion of geometry of this infinite-dimensional space and building configuration space quantum physics. This physics is deceptively simple at the level of principle, which can be expressed in a nutshell as the statement, "Do not quantize!": quantum physics is just classical physics in the world of classical worlds.

Quantum Jump and Self

Actually I was exaggerating when I said that quantum theory reduces to infinite-dimensional classical physics. The notion of quantum jump is also needed and becomes distilled as the quintessence of quantum physics.

As a matter of fact, the attempt to understand how to construct the S-matrix believed to code the predictions of any quantum theory forced me to ponder the basic problems of quantum measurement theory. It was clear from the beginning that these problems were more or less equivalent with the questions, "What is consciousness?" and "What is free will?" The idea that the quantum jump represents the basic unit of conscious experience, "moment of consciousness," was almost unavoidable.

What was new, and resolved the basic paradox of quantum measurement theory, was that the quantum jump replaces the entire four-dimensional evolution of a physical system from big bang to infinite future with a new one. Our conscious existence represents continual re-creation of the world, also of our geometric past, whereas subjectively experienced past is of course something fixed.

For some time I thought that the quantum jump was something irreducible, but it turned out that already standard measurement theory implies that it has a complex anatomy.

By the fractality of consciousness, this anatomy has structural counterparts at the level of our everyday consciousness. As the first step, the unitary process occurs and creates maximally entangled, holistic quantum state. Then a process analogous to analysis occurs and leads eventually to a minimally entangled, maximally classical state with remaining bound-state entanglement serving as a correlate for the experience of understanding. The quantum jump could be compared to a creative brainstorm followed by analysis giving as an outcome understanding.

Physicists love variational principles and in the case of the quantum jump the variational principle is the Negentropy Maximization Principle, briefly NMP, which roughly says that the information content of conscious experience is maximized. The detailed formulation of this principle led to a profound revision of the information concept.

Observer is in the standard quantum theory an outsider: measurements affect the measured system but the theory is completely silent about the observer as a physical system. The notion of self makes the observer part of the universe. Very roughly, self is a system able to avoid generation of bound-state entanglement: at the moment this occurs consciousness is lost and self becomes a part of a larger self.

The basic prediction is an infinite self hierarchy with the universe at the top of the hierarchy. This hierarchy directly reflects the hierarchy of matter (elementary particles, nuclei, atoms, . . .). Self experiences its subselves as mental images. Also the sharing and fusion of mental images is predicted and makes possible a kind of shared information, implying what we call moral and conscience necessary for the stability of social structures. Consciousness would not be so private as usually believed.

The New View about Time and Energy

The interpretation of the quantum jump as a moment of consciousness forced a revision of beliefs about the nature of time. There are two times: the geometric time of physicists, which is just the fourth dimension, and the subjective, experienced time for which the chronon is a single quantum jump. These two times are, to a reasonable approximation, identical.

However, when one tries to understand what for instance memories are, assuming geometric and subjective time to be identical, one ends up with grave difficulties. In the new conceptual framework it suffices to store memories in the brain of the geometric past, and memory recall involves communications with the brain of the geometric past. Obviously this makes us four-dimensional beings and only the special character of sensory input creates the illusion that we are three-dimensional.

Energy and time are dually related in quantum theory. If the process of discovery would obey the rules of logic, I should have immediately realized that inertial mass (energy) and gravitational mass (energy) represent two nonequivalent notions, and only with certain assumptions holding true are they in close approximation under normal circumstances. Hence Einstein's Equivalence Principle equating the two energies is not universally true in the TGD universe.

One deep implication is that inertial energy can have also negative values wheras gravitational energy is positive. In particular, at cosmological scales the universe has a vanishing density of inertial energy, and one can say that the universe as a whole is a vacuum with respect to inertial energy and other analogous quantum numbers. This means that the cosmology is maximally predictive and one avoids the question "What were the initial values?" which plagues standard-physics-based cosmology and has led to the introduction of the somewhat vague anthropic principle.

The new view about energy and time has rather radical implications. There is no physical death since my four-dimensional body continues to exist in the geometric past, presumably well and alive. The entire four-dimensional cosmos is a living system. Our geometric past is alive. Also the geometric future is living, and highly advanced civilizations are presumably patiently waiting for us to become mature for a contact. We ourselves do this communication when we remember.

The communications are indeed possible. Negative energy photons propagating into the geometric past and time-reflected back as positive energy photons (note the complete analogy with ordinary reflection) provide the means of communication. Negative energy photons behave just as the phase-conjugate photons discovered in the seventies. Hence we even have the rudiments of the needed technology. Even more, in the TGD universe there is a hierarchy of preferred frequencies and corresponding binary codes. We might be at the verge of becoming members of the four-dimensional community.

In light of this, the Fermi paradox, created by the mysterious absence of evidence about the presence of extraterrestrial life, might be only apparent. We might be constantly sharing mental images with the civilizations of the geometric future and past. The claimed extraterrestrials might be also futuroterrestrials, that is the humankind of the future sharing mental images with us now. The mysterious crop circles might be messages from the humankind of the geometric future, perhaps sent in attempt to help us to take the decisive step.

In principle, negative energy signals also mean a freedom from the restrictions posed by the finite velocity of light. Remote sensing and remote control of the distant geometric past practically instantaneously would be possible technological applications of the time mirror mechanism. Remote metabolism, making possible an instantaneous gain of positive energy by sending negative energy to the geometric past, would be a further application. It seems that the basic distinction between animate and inanimate matter is that living matter applies these mechanisms routinely.

Classical Nondeterminism and the Double-Slit Experiment

If I should name one of the most characteristic and most puzzling features of TGD, it would be the nondeterminism of the classical dynamics determining the spacetime surfaces as analogs of Bohr orbits, not for an electron in an atom but for the entire universe in eight-dimensional embedding space. This nondeterminism distinguishes the classical worlds of TGD from the completely predictable clockwork worlds of standard classical physics.

The physical meaning of this nondeterminism has become gradually clear. Classical nondeterministic physics is a kind of representation for deeper nondeterministic quantum physics at the level of the world of classical worlds. Not only quantum states, but also sequences of quantum jumps representing the contents of consciousness, are represented at the level of spacetime. This representation is far from being faithful but makes possible what we can say about the contents of consciousness using language. The rest requires direct experiencing.

Only a few days before writing this, I learned about an ingenious experiment of a physicist Shahriar Afshar, which seems to destroy both the Copenhagen interpretation and the many-worlds interpretation, as well as all interpretations equivalent with them. What the experiment demonstrates is that, in Afshar's variant of double-slit experiments, wave and particle aspects of the photon seem to be realized simultaneously, in conflict with Bohr's complementarity principle. If the spacetime surface provides a representation for both the situation before the photons became detected and after it, the paradox disappears. This is possible only due to classical nondeterminism.

p-Adic Numbers and Generalization of Number Concept

TGD has also led to a generalization of number concept. About a decade ago, I started to strongly feel that p-adic number fields Rp, $p = 2, 3, 5, \ldots$, one for each prime, might be very important for TGD. This led via a sequence of tortuous sidetracks to a further generalization of the notion of spacetime, and even more, to a generalization of number concept without which the formulation of quantum theory does not seem to be possible.

Real number fields R and p-adic number fields Rp can be regarded as completions of rational numbers, having rational numbers in common. The idea is to fuse them to a single larger structure along common rationals. At the spacetime level this means that spacetime sheets can be both real and p-adic. p-Adic spacetime sheets are identified as a spacetime correlate for cognition and intention, the mind stuff of Descartes. One motivation for the identification is that the p-adic counterpart of physics is nondeterministic, and this fits nicely with the nondeterminism of imagination.

The p-adic notion of distance differs dramatically from the real one. What is infinitesimal p-adically is infinite in the real sense. This means that p-adic spacetime sheets representing intentions and thoughts have literally infinite size in the real sense. The only possible conclusion seems to be that cognition and intentionality are cosmic phenomena. Our cognitive body would have an infinite size and infinite duration so that cognitively we would be eternal souls. Brain and body would serve only as sensory receptors and builders of symbolic representations about sensory input, besides serving as motor instruments controlled by "self."

About half a decade ago, I reached a second generalization of number concept inspired by the consciousness theory. A generalization of the notion of infinite number by generalizing the notion of divisibility to infinite context was in question. The outcome was the notion of infinite prime. Amazingly, the construction of infinite primes is structurally equivalent with a repeated second quantization, a strange activity to which quantum field theorists have devoted themselves. The interpretation of this mathematical construct remained however open.

Only quite recently it became clear that infinite numbers could be very relevant for the understanding of mathematical cognition. The point is that one can construct endless series of numbers which are equivalent with number 1 in real sense but in varous p-adic senses differ from unity. Take an infinite rational $Q1$ and divide it by another infinite rational $Q2$ such that the ratio is in a real sense finite rational q and divide this rational by q. What you get is just 1 in the real sense but not in various p-adic senses.

This means that every spacetime point is infinitely structured since one can multiply coordinates by any unit obtained in this manner. This is in sharp contrast with the prevailing idea that spacetime points are completely devoid of structure. Amazingly, the resulting structure is so complex that it makes it possible for a single spacetime point to represent in its structure even the entire quantum state of the entire universe!

This would mean that the universe is an algebraic hologram in a very precise sense: Brahman-Atman identity would be realized in the basic structure of spacetime. This structure would be completely invisible at the level of real physics, and would most naturally serve as the spacetime correlate for mathematical cognition: each spacetime point would represent Platonia of mathematical ideas. This idea is not new. Leibniz postulated that spacetime points are monads reflecting the universe in their structure but the idea was forgotten for centuries by physicists.

TGD and Quantum Biology

Some of the most interesting applications of TGD are in biology.

Many-sheeted spacetime allows realization of the idea that living systems are macroscopic quantum systems. Besides atomic spacetime sheets, there is an entire hierarchy of larger and colder spacetime sheets allowing the presence of Bose-Einstein condensates, such as superconductors.

Topological light rays (or "massless extremals")—the flux quanta of the electric field, giving rise to electrets—and flux tubes seem to be the basic new structural elements relevant for the self-organization of living matter. The predicted classical long-range electroweak forces and the role of neutrinos in biology are in sharp contrast with what the standard model would predict, but would explain the mysterious chiral selection in living matter.

What came as a surprise is that in the TGD universe the biological body is not all that is needed to understand life. Topological field quantization means that physical systems have also electromagnetic identity—field body or magnetic body. This body takes the role of intentional agent representing the "self" with which we usually identify ourselves. Libet's findings about strange time delays of consciousness support the view that information from the brain is communicated to a magnetic body by EEG waves with wavelengths measured using Earth size as natural length unit. Our field body would be of astrophysical size.

The time mirror mechanism based on phase conjugate light making possible remote metabolism has evolved to a rather detailed level, and connections with the physics of nonliving matter begin to emerge. For instance, the pioneering experiments of Nimtz about superluminal light velocities could be interpreted in terms of remote metabolism exercised by photon detectors using photons as their "food"! Energy metabolim is certainly one of the fundamental aspects of what it is to be alive.

With the help of several puzzling findings about genes, TGD leads to rather concrete ideas about the interpretation of DNA. Also a model for pre-biotic evolution emerges. For example, plasmoids, which I identify as magnetic structures containing plasma phase, are excellent candidates for primitive life forms. To my pleasant surprise, I learned some time ago that plasmoids have been found by Romanian physicists Lozneanu and Sanduloviciu to exhibit many of those aspects assigned to living matter.

These ideas are just those which are occupying my mind just now, and the model of animate matter involves also other building blocks which would deserve emphasis.

Feeedback Is Important

I have chosen to make TGD freely available to scientists and anyone interested in modern physics. Please feel free to read about my work at www.physics.helsinki.fi/~matpitka/. Besides the four books, the page contains articles about TGD. There are also tutor pages, but unfortunately these are out of date and remain so due to lack of time and resources. I am happy for feedback from physicists, biologists, neuroscientists, and other scientists who experience resonance with TGD. The feedback has indeed served as the key stimulator of ideas, and I am grateful for constructive criticism and have a passion for anomalies.

Matti Pitkänen
September 2004
Helsinki

Bibliography

NOTE: Internet references with the following prefixes are located at the website http://xxx.lanl.gov, the e-print archive site operated by Cornell University:

astro-ph: Astrophysics
cond-mat: Condensed Matter
gr-qc: General Relativity and Quantum Cosmology
hep-ex: High Energy Physics—Experiment
hep-ph: High Energy Physics—Phenomenology
hep-th: High Energy Physics—Theory
math.AG: Mathematics—Algebraic Geometry
math.DS: Mathematics—Dynamical Systems
math-ph: Mathematical Physics
nlin.CD: Nonlinear Sciences—Chaotic Dynamics
nucl-th: Nuclear Theory
physics: Physics
quant-ph: Quantum Physics.

***

Abanov, A. G. and P. B. Wiegmann. "Tunneling in the topological mechanism of superconductivity." cond-mat/9703157 (March 13, 1998).

Abbott, Edwin A. *Flatland: A Romance of Many Dimensions*. New York: Penguin Books, 1998.

Abbott, Richard B., Stephen M. Barr, and Stephen D. Ellis. "Kaluza-

Klein cosmologies and inflation." *Physical Review D* 30 (August 15, 1984): 720-27.

Albeverio, Sergio, Andrei Khrennikov, and Brunello Tirozzi. "*p*-Adic Dynamical Systems and Neural Networks." *Mathematical Models and Methods in Applied Sciences* 9 (December 1999): 1417-37.

Albeverio, S., P. E. Kloeden, and A. Khrennikov. "Human Memory as a *p*-Adic Dynamic System." *Theoretical and Mathematical Physics* 117 (1998): 1414-22.

Albrecht, Andreas and João Magueijo. "Time varying speed of light as a solution to cosmological puzzles." *Physical Review D* 59 (January 28, 1999): 043516 1-13.

Alhassid, Y. "The Statistical Theory of Quantum Dots." cond-mat/0102268 (February 15, 2001).

Ali, Syed Mustafa and Robert M. Zimmer. "Beyond Substance and Process: A New Framework for Emergence." In *Toward a Science of Consciousness II: The Second Tucson Discussions and Debates*, edited by Stuart R. Hameroff, Alfred W. Kaszniak, and Alwyn C. Scott, 585-92. Cambridge, Mass.: M. I. T. Press, 1998.

Altaisky, M. V. and B. G. Sidharth. "*p*-Adic physics below and above Planck scales." gr-qc/9802034 (February 16, 1998).

Alonso, D., J. G. Muga, and R. Sala Mayato. "Comment on 'Foundations of quantum mechanics: Connection with stochastic processes.'" *Physical Review A* 64 (June 8, 2001): 016101 1-3.

Ammosov, Vladimir and Guennadi Volkov. "Can Neutrinos Probe Extra Dimensions?" hep-ph/0008032 (August 3, 2000).

Anastassov, A. H. "Determination of the Fine Structure Constant by ." physics/9712044 (December 22, 1997).

Antoniadis, I. "A possible new dimension at a few TeV." *Physics Letters B* 246 (August 30, 1990): 377-84.

Antoniadis, Ignatios, Nima Arkani-Hamed, Savas Dimopoulos, and Gia Dvali. "New dimensions at a millimeter to a fermi and superstrings at a TeV." *Physics Letters B* 436 (September 24, 1998): 257-263.

Appelquist, Thomas, Hsin-Chia Cheng, and Bogdan A. Dobrescu. "Bounds on universal extra dimensions." *Physical Review D* 64 (June 21, 2001): 035002 1-10.

Argyres, Philip C., Savas Dimopoulos, and John March-Russell. "Black holes and sub-millimeter dimensions." *Physics Letters B* 441 (November 26, 1998): 96-104.

Argyris, J., C. I. Ciubotariu, and W. E. Weingaertner. "Fractal space signatures in quantum physics and cosmology—Space, time, matter, fields and gravitation." *Chaos, Solitons and Fractals* 11 (2000): 1671-1719.

Arkani-Hamed, Nima, Savas Dimopoulos, and Gia Dvali. "The hierarchy problem and new dimensions at a millimeter." *Physics Letters B* 429 (June 18, 1998): 263-72.

------. "Phenomenology, Astrophysics and Cosmology of Theories with Sub-Millimeter Dimensions and TeV Scale Quantum Gravity." hep-ph/9807344 (July 12, 1998).

Arkani-Hamed, Nima, Savas Dimopoulos, Gia Dvali, and Nemanja Kaloper. "Infinitely Large New Dimensions." *Physical Review Letters* 84 (January 24, 2000): 586-89.

------. "Manyfold Universe." hep-ph/9911386 (November 17, 1999).

Arkani-Hamed, Nima, Savas Dimopoulos, Nemanja Kaloper, and John March-Russell. "Early Inflation and Cosmology in Theories with Sub-Millimeter Dimensions." hep-ph/9903239 (April 16, 1999).

Armstrong, William. "The Bogdanov Brothers and their Hoax; or was it just Questionable Physics?" *PAM Bulletin* 30, no. 3, http://www.sla.org/division/dpam/pam~bulletin/vol30/no3/physics.html.

Asimov, Isaac. *The Neutrino: Ghost Particle of the Atom.* Garden City, N. Y.: Doubleday, 1966.

Audretsch, Jürgen and Michael Mensky. "Continuous Fuzzy Measurements and Visualization of a Quantum Transition." In *Quantum Future: From Volta and Como to the Present and Beyond; Proceedings of the Xth Max Born Symposium Held in Przesieka, Poland, 24-27 September 1997*, edited by Philippe Blanchard and Arkadiusz Jadczyk, 1-14. New York: Springer, 1999.

Auriemma, Giulio. "High energy neutrinos." astro-ph/0203331 (March 20, 2002).

Averin, D. V. "Quantum computing and quantum measurement with mesoscopic Josephson junctions." quant-ph/0008114 (August 27, 2000).

Avery, Samuel C. "Observers in Space and Time." *Physics Essays* 5 (1992): 185-89.

------. "The Theory of Dimensional Correspondence." *Physics Essays* 3 (1990): 133-46.

Avetisov, V. A., A. Kh. Bikulov, and V. A. Osipov. "*p*-Adic description of characteristic relaxation in complex systems." cond-mat/0210447 (October 21, 2002).

Baez, John. "The Bogdanov Affair." http://ntserv.fys.ku.dk/Presseklip/Klip/Bogdanov11112002.htm (November 11, 2002).

Barbour, Julian. *The End of Time: The Next Revolution in Physics.* New York: Oxford University Press, 1999.

Barceló, Carlos and Antonio Campos. "Braneworld physics from the analog-gravity perspective." hep-th/0206217 (June 24, 2002).

Barceló, Carlos, Stefano Liberati, and Matt Visser. "Analog gravity from Bose-Einstein condensates." gr-qc/0011026 (November 7, 2000).

Barrow, John D. and Hideo Kodama. "All Universes Great and Small." gr-qc/0105049 (May 15, 2001).

Barrow, John D. and David F. Mota. "Qualitative Analysis of Universes with Varying Alpha." gr-qc/0207012 (July 2, 2002).

Bass, L. "A Quantum Mechanical Mind-Body Interaction." *Foundations of Physics* 5 (1975): 159-72.

Battye, R. A., R. Crittenden, and J. Weller. "Cosmic concordance and the fine structure constant." astro-ph/0008265 (August 17, 2000).

Bearden, T. E. and W. Rosenthal. "On a Testable Unification of Electromagnetics, General Relativity, and Quantum Mechanics." *Proceedings of the 26th Intersociety Energy Conversion Engineering Conference* 4 (1991): 487-92.

Beck, Friedrich and John C. Eccles. "Quantum aspects of brain activity and the role of consciousness." *Proceedings of the National Academy of Science, USA* 89 (December 1992): 11357-61.

Berezin, Victor. "Quantum black hole: What is that?" *Nuclear Physics B (Proceedings Supplement)* 88 (2000): 34-39.

Bhattacharya, Kaushik and Palash B. Pal. "Neutrinos and magnetic fields: a short review." hep-ph/0212118 (March 13, 2003).

Blanchard, Philippe and Arkadiusz Jadczyk, eds. *Quantum Future: From Volta and Como to the Present and Beyond; Proceedings of the Xth Max Born Symposium Held in Przesieka, Poland, 24-27 September 1997.* New York: Springer, 1999.

Bob, P. and J. Faber. "Quantum Information in Brain Neural Networks and Electroencephalogram." *Neural Network World* 9 (1999): 365-72.

Bohm, David. *Wholeness and the Implicate Order.* London: Routledge and Kegan Paul, 1980.

Botta Cantcheff, Marcelo. "Generalized geometries and kinematics for Quantum Gravity." hep-th/0012036 (December 7, 2000).

------. "The 'phenomenological' space time and quantization." gr-qc/9804084 (April 30, 1998).

------. "Unified field theory from one-particle physics." hep-th/0012088 (December 11, 2000).

Brekke, Lee and Peter G. O. Freund. "*p*-Adic numbers in physics." *Physics Reports* 233 (October, 1993): 1-66.

Burgess, C.P., M. Majumdar, D. Nolte, F. Quevedo, G. Rajesh, and R.-J. Zhang. "The Inflatioary Brane-Antibrane Universe." hep-th/0105204 (August 27, 2001).

Burioni, R., D. Cassi, I. Meccoli, M. Rasetti, S. Regina, P. Sodano, and A. Vezzani. "Bose-Einstein condensation in inhomogeneous Josephson arrays." cond-mat/0004100 (November 14, 2000).

Caban, Pawel, Jakub Rembieliński, Kordian A. Smoliński, and Zbigniew Walczak. "Oscillations do not distinguish between massive and tachyonic neutrinos." hep-ph/0304221 (April 29, 2003).

Caldwell, D. O., R. N. Mohapatra, and S. J. Yellin. "Neutrinos, Large Extra Dimensions and Solar Neutrino Puzzle." hep-ph/0101043 (January 5, 2001).

Calvino, Italo. *Cosmicomics.* New York: Harcourt Brace Jovanovich, 1968.

Camp, Paul J. and John L. Safko. "Quantization Conditions in Curved Spacetime and Uncertainty-Driven Inflation." *International Journal of Theoretical Physics* 39 (2000): 1643-68.

Capra, Fritjof. *The Tao of Physics: An Exploration of the Parallels Between Modern Physics and Eastern Mysticism.* Boston: Bantam Books, 1975.

Carmichael, H. J. "Quantum Jumps Revisited: An Overview of Quantum Trajectory Theory." In *Quantum Future: From Volta and Como to the Present and Beyond; Proceedings of the Xth Max Born Symposium Held in Przesieka, Poland, 24-27 September 1997*, edited by Philippe Blanchard and Arkadiusz Jadczyk, 15-36. New York: Springer, 1999.

Carlip, S. "Quantum gravity: a progress report." *Reports on Progress in Physics* 64 (2001) 885-942.

Carugno, Enrico, Marco Litterio, Franco Occhionero, and Giuseppe Pollifrone. "Inflation in multidimensional quantum cosmology." *Physical Review D* 53 (June 15, 1996): 6863-74.

Casadio, Roberto and Benjamin Harms. "Black hole evaporation and compact extra dimensions." *Physical Review D* 64 (June 20, 2001): 024016 1-8.

------. "Black hole evaporation and large extra dimensions." *Physics Letters B* 487 (August 17, 2000): 209-14.

Casey, Terence W. "Prime Number Theory and the Fine-Structure Constant." *Physics Essays* 5 (1992): 345-46.

Castellani, Elena. "Reductionism, Emergence, and Effective Field Theories." physics/0101039 (January 6, 2001).

Castro, Carlos. "Fractal strings as an alternative justification for El Naschie's cantorian spacetime and the fine structure constant." *Chaos, Solitons & Fractals* 14 (2002): 1341-51.

------. "Fractal Strings as the Basis of Cantorian-Fractal Spacetime and the Fine Structure Constant." hep-th/0203086 (April 30, 2002).

------. "Is quantum space-time infinite dimensional." *Chaos, Solitons & Fractals* 11 (2000): 1663-70.

------. "A note on transfinite M theory and the fine structure constant." *Chaos, Solitons & Fractals* 14 (2002): 613-18.

Cerdonio, Massimo. "The search for gravitational waves." *Nuclear Physics B (Proceedings Supplement)* 88 (2000): 40-48.

Chamblin, Andrew, Csaba Csáki, Joshua Erlich, and Timothy J. Hollowood. "Black diamonds at brane junctions." *Physical Review D* 62 (July 18, 2000): 044012 1-7.

Chechelnitsky, A. M. "Mystery of the 'Magic Number' 137: Wave Genesis, Theoretical Representation, Role in the Universe." physics/0011035 (November 16, 2000).

Chen, Zeng-Bing and Yong-De Zhang. "Possible realization of Josephson charge qubits in two coupled Bose-Einstein condensates." cond-mat/0107368 (July 17, 2001).

Cheng, Hsin-Chia, Jonathan L. Feng, and Konstantin T. Matchev. "Kaluza-Klein Dark Matter." hep-ph/0207125 (July 10, 2002).

Chiao, Raymond Y. "Conceptual tensions between quantum mechanics and general relativity: Are there experimental consequences, e.g., superconducting transducers between electromagnetic and gravitational radiation?" gr-qc/0208024 (September 2, 2002).

------. "Tunneling Times and Superluminality: a Tutorial." quant-ph/9811019 (September 2, 2002).

Chistyakov, D. "Fractal Geometry for Images of Continuous Map of p-Adic Numbers and p-Adic Solenoids Into Euclidean Spaces." math.DS/0202089 (February 10, 2002).

Ćirković, Milan M. and Nick Bostrom. "Cosmological Constant and the Final Anthropic Hypothesis." *Astrophysics and Space Science* 274 (2000): 675-87.

Cole, K. C. *The Hole in the Universe: How Scientists Peered Over the Edge of Emptiness and Found Everything.* New York: Harcourt, 2001.

Conrad, Michael. "Anti-Entropy and the Origin of Initial Conditions." *Chaos, Solitons & Fractals* 7 (1996): 725-45.

Cornish, Neil J. and David N. Spergel. "A small universe after all?" *Physical Review D* 62 (September 26, 2000): 087304 1-4.

Corradini, Olindo, Alberto Iglesias, Zurab Kakushadze, and Peter Langfelder. "Gravity on a 3-brane in 6D bulk." *Physics Letters B* 521 (November 22, 2001): 96-104.

Davidson, Mark. "Tachyons, Quanta and Chaos." Palo Alto, Calif.: Spectel Research Corporation, mdavid@spectel.com (February 15, 2001).

Davies, P. C. W. and Julian Brown, eds. *Superstrings: A Theory of Everything.* Cambridge, U.K.: Cambridge University Press, 1988.

de Aquino, Fran. "Superparticles from the Initial Universe and Deduction of the Fine Structure Constant and Uncertainty Principle Directly from the Gravitation Theory." physics/0103093 (March 29, 2001).

de Carvalho, C. A. A. and R. M. Cavalcanti. "Tunnelings as Catastrophe." hep-th/9511005 (November 3, 1995).

Deffayet, Cédric and Gia Dvali. "Accelerated universe from gravity leaking to extra dimensions." *Physical Review D* 65 (January 28, 2002): 044023 1-9.

Demir, D.A. "Stable Q-balls from extra dimensions." *Physics Letters B* 495 (December 14, 2000): 357-62

de Oliveira, M. C. "Teleportation of a Bose-Einstein condensate state by controlled elastic collisions." quant-ph/0212118 (December 19, 2002).

Derbyshire, John. *Prime Obsession: Bernard Riemann and the Greatest Unsolved Problem in Mathematics.* Washington, D. C.: Joseph Henry Press, 2003.

de Vries, Andreas and Theodor Schmidt-Kaler. "The black hole tunnel phenomenon." gr-qc/0112069 (December 26, 2001).

Dienes, Keith R. "Shape versus Volume: Making Large Flat Extra Dimensions Invisible." *Physical Review Letters* 88 (January 7, 2002): 011601 1-4.

Dienes, Keith R., Emilian Dudas, and Tony Gherghetta. "Extra spacetime dimensions and unification." *Physics Letters B* 436 (September 17, 1998): 55-65.

------. "Grand unification at intermediate mass scales through extra dimensions." *Nuclear Physics B* 537 (1999): 47-108.

------. "TeV-Scale GUTs." hep-ph/9807522 (July 28, 1998).

Dighe, Amol S. and Anjan S. Joshipura. "Neutrino anomalies and large extra dimensions." hep-ph/0105288 (May 28, 2001).

Djordjevic, G. S., B. Dragovich, and Lj. D. Nesic. "*p*-Adic Quantum Cosmology." *Nuclear Physics B, Proceedings Supplements* 104 (January, 2002): 197-200

Djordjevic, G. S., B. Dragovich, Lj. D. Nesic, and I. V. Volovich. "p-Adic and Adelic Minisuperspace Quantum Cosmology." gr-qc/0105050 (April 2, 2002).

Dolgov, A. D. "Neutrinos in cosmology." hep-ph/0202122 (April 19, 2002).

Donald, M. J. "Quantum Theory and the Brain." *Proceedings of the Royal Society of London, Series A, Mathematical and Physical Sciences* 427 (January 8, 1990): 43-93.

Dragovich, Branko. "Non-Archimedean Geometry and Physics on Adelic Spaces." math-ph/0306023 (June 9, 2003).

Dragovich, Branko and Ljubisa Nesic. "On p-Adic Numbers in Gravity." *Balkan Physics Letters* 6 (April 1, 1998): 78-81.

------. "p-Adic and Adelic Generalization of Quantum Cosmology." *Gravitation & Cosmology* 5 (1999): 222-28.

Drake, Frank and Dava Sobel. *Is Anyone Out There? The Scientific Search for Extraterrestrial Intelligence.* New York: Dell Publishing, 1992.

Dubischar, Daniel, Volker Matthias Gundlach, Andrei Khrennikov, and Oliver Steinkamp. "Attractors of random dynamical systems over p-adic numbers and a model of 'noisy' cognitive processes." *Physica D* 130 (1999): 1-12.

Durrer, Ruth. "Frontiers of the Universe: What do we know, what do we understand?" astro-ph/0205101 (May 7, 2002).

Durrer, Ruth and Joachim Laukenmann. "The oscillating universe: an alternative to inflation." *Classical and Quantum Gravity* 13 (1996): 1069-87.

Dvali, G. "Submillimeter Extra Dimensions and TeV-Scale Quantum Gravity." *International Journal of Theoretical Physics* 39 (2000): 1717-29.

Dvali, G., G. Gabadadze and M. Porrati. "Metastable gravitons and infinite volume extra dimensions." *Physics Leters B* 484 (June 29, 2000): 112-18.

Dvali, Gia and Alexei Yu. Smirnov. "Probing Large Extra Dimensions with Neutrinos." hep-ph/9904211 (June 15, 1999).

Dymnikova, Irina and Maxim Khlopov. "Decay of cosmological constant as Bose condensate evaporation." astro-ph/0102094 (February 15, 2001).

Dzhunushaliev, V. D. "Domain with Noncompactified Extra Dimensions in the Multidimensional Universe with Compactified Extra Dimensions." *General Relativity and Gravitation* 30 (1998): 1655-61.

Dziarmaga, J., A. Smerzi, W. H. Zurek, and A. R. Bishop. "Dynamics of Quantum Phase Transition in an Array of Josephson Junctions." cond-mat/0110627 (October 30, 2001).

Easther, Richard, Brian R. Greene, William H. Kinney, and Gary Shiu. "Inflation as a Probe of Short Distance Physics." hep-th/0104102 (May 18, 2001).

Eccles, John C. "Do Mental Events Cause Neural Events Analogously to the Probability Fields of Quantum Mechanics?" *Proceedings of the Royal Society of London, Series B, Biological Sciences* 227 (May 22, 1986): 411-28.

------. "Evolution of consciousness." *Proceedings of the National Academy of Science, USA* 89 (August 1992): 7320-24.

------. *Evolution of the Brain: Creation of the Self.* New York: Routledge, 1989.

Eccles, John C. and Otto Creutzfeldt, eds. *The Principles of Design and Operation of the Brain.* New York: Springer-Verlag, 1990.

Eguchi, K. et al. "First Results from KamLAND: Evidence for Reactor Antineutrino Disappearance." *Physical Review Letters* 90 (January 17, 2003): 021802 1-6.

Ehresmann, A. C. and J.-P. Vanbremeersch. "Emergence Processes Up to Consciousness Using the Multiplicity Principle and Quantum Physics." *AIP Conference Proceedings*, no. 627 (2002): 221-33.

Ellis, John. "Particle Candidates for Dark Matter." astro-ph/9812211 (December 10, 1998).

Emparan, Roberto, Gary T. Horowitz, and Robert C. Myers. "Black Holes Radiate Mainly on the Brane." *Physical Review Letters* 85 (July 17, 2000): 499-502.

Emparan, Roberto, Manuel Masip, and Riccardo Rattazzi. "Cosmic rays as probes of large extra dimensions and TeV gravity." *Physical Review D* 65 (February 27, 2002): 064023 1-7.

Fairbairn, Malcolm and Michel H.G. Tytgat. "Inflation from a Tachyon Fluid?" hep-th/0204070 (June 26, 2002).

Faisal, F. H. M. "Quantum Chaos, Algorithmic Paradigm and Irreversibility." In *Quantum Future: From Volta and Como to the Present and Beyond; Proceedings of the Xth Max Born Symposium Held in Przesieka, Poland, 24-27 September 1997*, edited by Philippe Blanchard and Arkadiusz Jadczyk, 47-57. New York: Springer, 1999.

Feng, Jonathan L. and Alfred D. Shapere. "Black Hole Production by Cosmic Rays." *Physical Review Letters* 88 (January 14, 2002): 021303 1-4.

Ferguson, William. "Books in Brief." *New York Times Book Review,* January 12, 2003, 19.

Ferris, Timothy. *The Mind's Sky: Human Intelligence in a Cosmic Context.* New York: Bantam Books, 1992.

Flanagan, É. É., S.-H. H. Tye, and I. Wasserman. "Cosmology of the brane world." *Physical Review D* 62 (June 21, 2000): 024011 1-20.

Frampton, Paul H., P. Q. Hung, and Marc Sher. "Quarks and Leptons Beyond the Third Generation." *Physics Reports* 330 (2000): 263-348.

French, A. P. *Special Relativity*. New York: W. W. Norton, 1968.

Gamwell, Lynn. *Exploring the Invisible: Art, Science, and the Spiritual*. Princeton, N. J.: Princeton University Press, 2002.

Gando, Y. et al. "Search for e from the Sun at Super-Kamiokande-I." *Physical Review Letters* 90 (May 1, 2003): 171302 1-5.

García-Calderón, Gastón and Jorge Villavicencio. "Early times in tunneling." quant-ph/0008014 (August 3, 2000).

Gardner, Carl L. "Primordial inflation and present-day cosmological constant from extra dimensions." *Physics Leters B* 524 (January 3, 2002): 21-25.

Ghoshal, Debashis and Ashoke Sen. "Tachyon Condensation and Brane Descent Relations in p-Adic String Theory." hep-th/0003278 (March 30, 2000).

Gibbons, G. W. "Cosmological Evolution of the Rolling Tachyon." hep-th/0204008 (April 18, 2002).

Giddings, Steven B. and Scott Thomas. "High energy colliders as black hole factories: The end of short distance physics." *Physical Review D* 65 (2002): 056010 1-12.

Giovanazzi, S., A. Smerzi, and S. Fantoni. "Josephson effects in dilute Bose-Einstein condensates." physics/0002041 (February 21, 2000).

Giudice, Gian F., Riccardo Rattazzi, and James D. Wells. "Quantum gravity and extra dimensions at high-energy colliders." *Nuclear Physics B* 544 (1999): 3-38.

Glashow, Sheldon L. *The Charm of Physics*. New York: Touchstone, 1991.

González-Díaz, Pedro F. "Observable effects from spacetime tunneling." gr-qc/9708044 (August 19, 1997).

Greene, Brian. *The Elegant Universe: Superstrings, Hidden Dimensions, and the Quest for the Ultimate Theory*. New York: W. W. Norton & Company, 1999.

Gregory, Ruth, Valery A. Rubakov, and Sergei M. Sibiryakov. "Opening up Extra Dimensions at Ultralarge Scales." *Physical Review Letters* 84 (June 26, 2000): 5928-31

Gribbin, John. *Q Is for Quantum: An Encyclopedia of Particle Physics*. New York: Simon and Schuster, 1998.

Grosse, Harald and Karl-Georg Schlesinger. "Spinfoam models for M-theory." *Physics Letters B* 528 (2002): 106-110.

Grössing, Gerhad. *Quantum Cybernetics: Toward a Unification of Relativity and Quantum Theory via Circularly Causal Modeling.* New York: Springer-Verlag, 2000.

Grossman, Yuval, Alexander L. Kagan, and Matthias Neubert. "Trojan Penguins and Isospin Violation in Hadronic B Decays." hep-ph/9909297 (November 2, 1999).

Gudmundsson, Einar H. and Gunnlaugur Björnsson. "Dark Energy and the Observable Universe." *The Astrophysical Journal* 565 (January 20, 2002): 1-16.

Guendelman, E. I. and David A. Owen. "Universe Creation, Entropy, and Extra Dimensions." *General Relativity and Gravitation* 21 (1989): 201-10.

Gupta, S., K. Dieckmann, Z. Hadzibabic, and D. E. Pritchard. "Contrast Interferometry Using Bose-Einstein Condensates to Measure h/m and the Fine Structure Constant." cond-mat/0202452 (July 8, 2002).

Gurin, V. S. and A. P. Trofimenko. "Higher-dimensional white holes." *Pramana—Journal of Physics* 36 (May, 1991): 511-18

------. "White Holes in Kaluza-Klein Theory: Windows from Higher Dimensions." *Physics Letters B* 241 (May 17, 1990): 328-31.

Guth, Alan. "Inflationary universe: A possible solution to the horizon and flatness problems." *Physical Review D* 23 (January 15, 1981): 347-56.

------. *The Inflationary Universe: The Quest for a New Theory of Cosmic Origins.* Reading, Mass.: Addison-Wesley, 1997.

Hameroff, Stuart R. "Did Consciousness Cause the Cambrian Evolutionary Explosion?" In *Toward a Science of Consciousness II: The Second Tucson Discussions and Debates,* edited by Stuart R. Hameroff, Alfred W. Kaszniak, and Alwyn C. Scott, 421-37. Cambridge, Mass.: M. I. T. Press, 1998.

Hameroff, Stuart R., Alfred W. Kaszniak, and David J.Chalmers, eds. *Toward a Science of Consciousness III: The Third Tucson Discussions and Debates.* Cambridge, Mass.: M. I. T. Press, 1999.

Hameroff, Stuart R., Alfred W. Kaszniak, and Alwyn C. Scott, eds. *Toward a Science of Consciousness: The First Tucson Discussions and Debates.* Cambridge, Mass.: M. I. T. Press, 1996.

------. *Toward a Science of Consciousness II: The Second Tucson Discussions and Debates.* Cambridge, Mass.: M. I. T. Press, 1998.

Hameroff, Stuart R. and Roger Penrose. "Orchestrated Reduction of Quantum Coherence in Brain Microtubules: A Model for Consciousness." In *Toward a Science of Consciousness: The First Tucson Discussions and Debates,* edited by Stuart R.

Hameroff, Alfred W. Kaszniak, and Alwyn C. Scott, 507-540. Cambridge, Mass.: M. I. T. Press, 1996.

Hawking, Stephen. *Black Holes and Baby Universes and other Essays.* New York: Bantam Books, 1993.

------. *A Brief History of Time: The Updated and Expanded Tenth Anniversary Edition.* New York: Bantam Books, 1996.

------. *The Theory of Everything: The Origin and Fate of the Universe.* Beverly Hills: New Millennium Press, 2002.

------. *The Universe in a Nutshell.* New York: Bantam Books, 2001.

Hawking, Stephen and Roger Penrose. *The Nature of Space and Time.* Princeton, N. J.: Princeton University Press, 1996.

Haxton, W. C. "Nuclear Problems in Astrophysics." nucl-th/0301005 (January 3, 2003).

Hazen, Robert. "The Great Unanswered Questions." *Technology Review* 100 (July, 1997): 28.

Hepp, K. "Toward the Demolition of a Computational Quantum Brain." In *Quantum Future: From Volta and Como to the Present and Beyond; Proceedings of the Xth Max Born Symposium Held in Przesieka, Poland, 24-27 September 1997,* edited by Philippe Blanchard and Arkadiusz Jadczyk, 92-104. New York: Springer, 1999.

Herbert, Nick. *Quantum Reality: Beyond the New Physics.* New York: Doubleday, 1985.

Hofstadter, Douglas R. *Gödel, Escher, Bach: An Eternal Golden Braid.* New York: Vintage Books, 1979.

Hollands, Stefan and Robert M. Wald. "An Alternative to Inflation." gr-qc/0205058 (May 31, 2002).

Horwitz, Gerald. "Time and Entropy from Semi-classical Tunneling of the Cosmological Scale Function." gr-qc/9507058 (August 7, 1995).

Horwitz, L. P., J. Levitan, and Y. Ashkenazy. "Complexity, Tunneling and Geometrical Symmetry." cond-mat/9604130 (January 1, 2003).

Hosoya, A., A. Carlini, and T. Shimomura. "Generalized second law of black hole thermodynamics and quantum information theory." *Physical Review D* 63: 104008 1-5 (April 9, 2001).

Huning, H. "Cognition and Neural Network Modelling." In *Artificial Neural Networks: Proceedings of the 1992 International Conference (ICANN-92)* 2: 1305-9. Amsterdam: Elsevier, 1992.

Inoue, Kaiki Taro and Naoshi Sugiyama. "How large is our universe?" astro-ph/0205394 (May 23, 2002).

Insinna, E. M. "Synchronicity and Emergent Nonlocal Information in Quantum Systems." In *Toward a Science of Consciousness:*

The First Tucson Discussions and Debates, edited by Stuart R. Hameroff, Alfred W. Kaszniak, and Alwyn C. Scott, 597-608. Cambridge, Mass.: M. I. T. Press, 1996.

Jadczyk, Arkadiusz and Laura Knight-Jadczyk. "The Bogdanov Singularity: The Bogdanoff Brothers in the Eye of a Cyclone." *Quantum Future Physics*, http://cass.eahosting. com/cass/bog-lanoy_trans.htm (2003).

Jaffe, K. and C. Fonck. "Energetics of Social Phenomena: Physics Applied to Evolutionary Biology." *Nuovo Cimento D* 16D (June 1994): 543-53

Jaynes, Julian. *The Origin of Consciousness in the Breakdown of the Bicameral Mind*. Boston: Houghton Mifflin Company, 1976.

Jibu, M. and K. Yasue. "What Is Mind? Quantum Field Theory of Evanescent Photons in Brain as Quantum Theory of Consciousness." *Informatica* 21 (September 1997): 471-90.

Johnson, George. "In Theory, It's True (Or Not)." *New York Times*, November 17, 2002, sec. 1: 4.

Jonas, Gerald. "Science Fiction." *New York Times Book Review*, November 10, 2002, 21.

Jones, Nicholas, Horace Stoica, and S.-H. Henry Tye. "Brane Interaction as the Origin of Inflation." hep-th/0203163 (March 21, 2002).

Kafatos, Menas and Robert Nadeau. *The Conscious Universe: Parts and Wholes in Physical Reality*. New York: Springer-Verlag, 2000.

Kaku, Michio. *Hyperspace: A Scientific Odyssey through Parallel Universes, Time Warps, and the 10th Dimension*. New York: Oxford University Press, 1994.

Kane, Gordon. *The Particle Garden: Our Universe as Understood by Particle Physicists*. Reading, Mass.: Addison-Wesley, 1995.

Kane, Gordon. "Experimental Evidence for More Dimensions." *Physics Today*: 51 (May 1998): 13-15.

Karlin, Susan. "A Sculptor Works Up an Exposé of the Stars' Secrets." *New York Times*, November 3, 2002, Arts & Entertainment: 21, 28.

Kent, Adrian. "Night Thoughts of a Quantum Physicist." physics/9906040 (February 8, 2000).

Keshavamurthy, Srihari. "Dynamical tunneling in molecules: role of the classical resonances and chaos." nlin.CD/0210051 (October 23, 2002).

Ketterle, W., D. S. Durfee, and D. M. Stamper-Kurn. "Making, probing and understanding Bose-Einstein condensates." cond-mat/9904034 (April 5, 1999).

Khoury, Justin, Burt A. Ovrut, Paul J. Steinhardt, and Neil Turok.

"Ekpyrotic universe: Colliding branes and the origin of the hot big bang." *Physical Review D* 64 (2001): 123522 1-24.

Khrennikov, Andrei. "Classical and quantum mechanics on information spaces with applications to cognitive, psychological, social and anomalous phenomena." quant-ph/0003016 (March 4, 2000).

------. "Information Interpretation of *p*-Adic Physics." *Mathematical Physics* 62 (2000): 117-19.

------. "*p*-Adic Quantum-Classical Analogue of the Heisenberg Uncertainty Relations." *Nuovo Cimento B* 112B (April 1997): 555-60

------. "*p*-Adic Stochastics with Applications to the Einstein-Podolsky-Rosen Paradox." In *Chaos—The Interplay Between Stochastic and Deterministic Behaviour: Proceedings of the 31st Winter School of Theoretical Physics*, 457-60. Berlin: Springer-Verlag, 1995.

------. "Quantum-like formalism for cognitive measurements." quant-ph/0111006 (February 16, 2003).

------. "Real Non-Archimedean Structure of Spacetime." *Teoreticheskaya i Matematicheskaya Fizika* 86 (February 1991): 177-90.

------. "Statistical Interpretation of *p*-Adic Quantum Theories with *p*-Adic Valued Wave Functions." *Journal of Mathematical Physics* 36 (December, 1995): 6625-32.

Kiefer, Claus and Erich Joos. "Decoherence: Concepts and Examples." In *Quantum Future: From Volta and Como to the Present and Beyond; Proceedings of the Xth Max Born Symposium Held in Przesieka, Poland, 24-27 September 1997*, edited by Philippe Blanchard and Arkadiusz Jadczyk, 105-28. New York: Springer, 1999.

Kirn, Walter. "Winging It: They also fight, who fly." *New York Times Magazine*, January 18, 2003, 11-12.

Kitcher, Philip. "The Mind Mystery." *New York Times Book Review*, February 4, 1990, 20.

Klipa, N. and S. D. Bosanac. "Understanding quantum effects from classical principles." quant-ph/0010089 (October 26, 2000).

Kofman, Lev and Andrei Linde. "Problems with Tachyon Inflation." hep-th/0205121 (May 14, 2002).

Kragh, Helge. *Quantum Generations: A History of Physics in the Twentieth Century.* Princeton, N. J.: Princeton University Press, 1999.

Krauss, Lawrence and Glenn D. Starkman. "Life, the Universe, and Nothing: Life and Death in an Ever-Expanding Universe." *The Astrophysical Journal* 531 (March 1, 2000): 22-30.

Kröger. Helmut. "Nonlinear Dynamics in Quantum Physics—

Quantum Chaos and Quantum Instantons." *Proceedings of the Tenth International Conference on Computation and Applied Mathematics,* quant-ph/0302169 (February 21, 2003).

Kumar, Pradeep, Plamen Ch. Ivanov, and H. Eugene Stanley. "Information Entropy and Correlations in Prime Numbers." cond-mat/0303110 (April 8, 2003).

Kwon, Jang-Hee. "The Structural Changes in the Relations of South and North Korea: Nonequilibrium Thermodynamics Approach." In *Systems Thinking, Globalization of Knowledge, and Communitarian Ethics: Proceedings of the Forty-First Annual Meeting of the International Society for the Systems Sciences,* 655-68. Louisville, Ky.: International Society for the Systems Sciences, 1997.

Laurikainen, K. V. "Atoms and Consciousness as Complementary Elements of Reality." *European Journal of Physics* 11 (March, 1990): 65-74.

Leblond, Frédéric. "Geometry of large extra dimensions versus graviton emission." *Physical Review D* 64 (July 27, 2001): 045016 1-11.

Lederman, Leon and Dick Teresi. *The God Particle: If the Universe Is the Answer, What Is the Question?* Boston: Houghton Mifflin, 1993.

Lee, Jinyoul, B. Jayatilaka, and Ron Chi-Wai Kwok. "Virtual Organization: Duality of Human Identities in Consciousness and Entity." *Issues and Trends of Information Technology Management in Contemporary Organizations* 1, part 1: 427-28.

Le Shan, Lawrence and Henry Margenau. *Einstein's Space and Van Gogh's Sky.* New York: Macmillan, 1982.

Lessing, Doris. *The Four-Gated City.* New York: New American Library, 1976.

Levin, Janna. "Topology and the Cosmic Microwave Background." gr-qc/0108043 (August 20, 2001).

Levin, Janna J. "Inflation from extra dimensions." *Physics Letters B* 343 (January 19, 1995): 69-75.

Levin, Janna J. and Katherine Freese. "Possible solution to the horizon problem: Modified aging in massless scalar theories of gravity." *Physical Review D* 47 (May 15, 1993): 4282-90.

Lewenstein, M., D. Bruß, J. I. Cirac, B. Kraus, M. Kus, J. Samsonowicz, A. Sanpera, and R. Tarrach. "Separability and distillability in composite quantum systems—a primer." *Journal of Modern Optics* 47 (2000): 2481-99.

Li, Li-Xin and J. Richard Gott, III. "Inflation in Kaluza-Klein Theory: Relation Between the Fine-Structure Constant and the

Cosmological Constant." astro-ph/9804311 (April 28, 1998).

Li, Xin-zhou, Dao-jun Liu, and Jian-gang Hao. "On the tachyon inflation." hep-th/0207146 (July 16, 2002).

Lightman, Alan. *Einstein's Dreams.* New York: Pantheon Books, 1993.

Limongelli, C. and H. W. Loidl. "Rational Number Arithmetic by Parallel *p*-Adic Algorithms." In *Parallel Computation: Second International ACPC Conference Proceedings,* 72-86. Berlin: Springer-Verlag, 1993.

Ling, F.-S. "A minimal model with large extra dimensions to fit the neutrino data." hep-ph/0105186 (May 18, 2001).

Liu, Ying. "Neurons, Psychons, and Emotion." In *Applications and Science of Artificial Neural Networks,* edited by Steven K. Rogers and Dennis W. Ruck, 184-98. Proceedings SPIE—The International Society for Optical Engineering, 1995.

Lodge, David. *Consciousnes and the Novel: Connected Essays.* Cambridge, Mass.: Harvard University Press, 2002.

Lopez de Lema, J. "Two-Behavior Physics: On Matter, Vacuum, and Consciousness." *Physics Essays* 3 (December, 1990): 323-30.

Loss, Daniel and David P. DiVincenzo. "Quantum Computation with Quantum Dots." cond-mat/9701055 (July 20, 1997).

Lyre, Holger. "Against Measurement?—On the Concept of Information." In *Quantum Future: From Volta and Como to the Present and Beyond; Proceedings of the Xth Max Born Symposium Held in Przesieka, Poland, 24-27 September 1997,* edited by Philippe Blanchard and Arkadiusz Jadczyk, 139-49. New York: Springer, 1999.

Maia, M. D. and V. Silveira. "Window to extra dimensions near a black hole." *Physical Review D* 48 (July 15, 1993): 954-57.

Majerník, Vladimír. "A Geometrical Description of Quantum Mechanics." *Physics Essays* 9 (1996): 419-428.

------. "Information Content in Quantum-Mechanical Systems." *Il Nuovo Cimento* LXIV A (December 1, 1969): 501-6.

Mak, M. K. and T. Harko. "Cosmological particle production in five-dimensional Kaluza-Klein theory." *Classical and Quantum Gravity* 16 (1999): 4085-99.

Malek, M. et al. "Search for Supernova Relic Neutrinos at Super-Kamiokande." *Physical Review Letters* 90 (February 13, 2003): 061101 1-5.

Marcer, P. J. "Getting Quantum Theory Off the Rocks: Nature as We Consciously Perceive It Is Quantum Reality!" In *14th International Congress on Cybernetics,* 435-40. Avon, U. K.: Aikido Enterprises, 1995.

Marshall, Ian. "Some phenomenological implications of a quantum model of consciousness." *Minds and Machines* 5 (November 1995): 609-20.

Marshall, Ian and Danah Zohar. *Who's Afraid of Schrödinger's Cat? All the New Science Ideas You Need to Keep Up with the New Thinking.* New York: William Morrow, 1997.

Matzke, Douglas J. "Consciousness: A New Computational Paradigm." In *Toward a Science of Consciousness: The First Tucson Discussions and Debates,* edited by Stuart R. Hameroff, Alfred W. Kaszniak, and Alwyn C. Scott, 569-77. Cambridge, Mass.: M. I. T. Press, 1996.

Mauceli, E., Z. K. Geng, W. O. Hamilton, W. W. Johnson, S. Merkowitz, A. Morse, B. Price, and N. Solomonson. "The Allegro gravitational wave detector: Data acquisition and analysis." *Physical Review D* 54 (July 15, 1996): 1264-75.

Mazumdar, Anupam, Sudhakar Panda, and Abdel Pérez-Lorenzana. "Assisted inflation via tachyon condensation." hep-ph/0107058 (August 15, 2001).

Mazur, Pawel O. and Emil Mottola. "Gravitational Condensate Stars: An Alternative to Black Holes." gr-qc/0109035 (February 27, 2002).

Mehra, Jagdish. *The Solvay Conferences on Physics: Aspects of the Development of Physics since 1911.* Boston: D. Reidel, 1975.

Mendelsohn, Daniel. "A Dance to the Music of Time." *New York Times Book Review,* January 26, 2003, 12.

Miković, A. "Spin foam models of matter coupled to gravity." *Classical and Quantum Gravity* 19 (2002): 2335-53.

Miller, Arthur I. *Einstein, Picasso: Space, Time, and the Beauty That Causes Havoc.* New York: Basic Books, 2001.

Minahan, Joseph A. "Mode Interactions of the Tachyon Condensate in p-Adic String Theory." hep-th/0102071 (February 13, 2001).

Miramonti, Lino and Franco Reseghetti. "Solar Neutrino Physics: historical evolution, present status and perspectives." hep-ex/0302035 (February 25, 2003).

Moeller, Nicolas and Martin Schnabl. "Tachyon condensation in open-closed p-adic string theory." hep-th/0304213 (April 25, 2003).

Moeller, Nicolas and Barton Zwiebach. "Dynamics with Infinitely Many Time Derivatives and Rolling Tachyons." hep-th/0207107 (September 4, 2002).

Mongan, T.R. "(N + 1)-Dimensional Quantum Mechanical Model for a Closed Universe." *International Journal of Theoretical Physics* 38 (1999): 1521-29.

Moore, M. G. and P. Meystre. "Generating entangled atom-photon pairs from Bose-Einstein condensates." quant-ph/0004083 (April 20, 2000).

Mould, Richard A. "Consciousness and Endogenous State Reduction: Two Experiments." *Foundations of Physics Letters* 14 (2001): 377-86.

------. "Consciousness and Quantum Mechanics." *Foundations of Physics* 28 (1998): 1703-18.

------. "Quantum Consciousness." *Foundations of Physics* 29 (1999): 1951-61.

Muschamp, Herbert. "A See-Through Library of Shifting Shapes and Colors." *New York Times,* January 19, 2003, 35.

Ne'eman, Yuval. "Quantizing gravity and spacetime: Where do we stand?" *Annalen Physik (Leipzig)* 8 (1999):. 3-17.

Ni, Guang-Jiong. "Evidence for Neutrino Being Likely a Superluminal Particle." hep-ph/0206296 (June 28, 2002).

Nimtz, Günter. "Superluminal Tunneling Devices." physics/0204043 (April 16, 2002).

Norris, Christopher. *Quantum Theory and the Flight from Realism: Philosophical Responses to Quantum Mechanics.* New York: Routledge, 2000.

Olavo, L. S. F. "Quantum Mechanics as a Classical Theory I: Non-relativistic Theory". quant-ph/9503020 (March 31, 1995).

------. "Quantum Mechanics as a Classical Theory II: Relativistic Theory". quant-ph/9503021 (March 31, 1995).

------. "Quantum Mechanics as a Classical Theory III: Epistemology." quant-ph/9503022 (March 31, 1995).

------. "Quantum Mechanics as a Classical Theory IV: The Negative Mass Conjecture." quant-ph/9503024 (March 31, 1995).

------. "Quantum Mechanics as a Classical Theory V: The Quantum Schwartzchild Problem." quant-ph/9503025 (March 31, 1995).

------. "Quantum Mechanics as a Classical Theory VI: The Classical Spin." quant-ph/9509012 (September 20, 1995).

------. "Quantum Mechanics as a Classical Theory VII: The Classical Spin Eigenfunctions." quant-ph/9509013 (September 20, 1995).

------. "Quantum Mechanics as a Classical Theory VIII: Second Quantization." quant-ph/9511028 (November 21, 1995).

------. "Quantum Mechanics as a Classical Theory IX: The Formation of Operators and Quantum Phase-Space Densities." quant-ph/9511039 (November 27, 1995).

------. "Quantum Mechanics as a Classical Theory X: Quantization in Generalized Coordinates." quant-ph/9601002 (January 3, 1996).

------. "Quantum Mechanics as a Classical Theory XI: Thermodynamics and Equilibrium." quant-ph/9607002 (July 1, 1996).

------. "Quantum Mechanics as a Classical Theory XII: Diffraction and Interference." quant-ph/9607003 (July 1, 1996).

------. "Quantum Mechanics as a Classical Theory XIII: The Tunnel Effect." quant-ph/9609003 (September 4, 1996).

------. "Quantum Mechanics as a Classical Theory XIV: Connection with Stochastic Processes." quant-ph/9609023 (September 27, 1996).

------. "Quantum Mechanics as a Classical Theory XV: Thermodynamical Derivation." quant-ph/9703006 (March 5, 1997).

------. "Quantum Mechanics as a Classical Theory XVI: Positive-Definite Densities." quant-ph/9704004 (April 2, 1997).

Olin, Dirk. "Ethnomathematics." *The New York Times Magazine*, February 23, 2003, 23-24.

Olinto, A. V. "Messengers of the Extreme Universe." astro-ph/0305177 (May 11, 2003).

Olive, Keith A. and Maxim Pospelov. "Evolution of the Fine Structure Constant Driven by Dark Matter and the Cosmological Constant." hep-ph/0110377 (October 29, 2001).

Omnés, Roland. "On Interpretation and Decoherence." In *Quantum Future: From Volta and Como to the Present and Beyond; Proceedings of the Xth Max Born Symposium Held in Przesieka, Poland, 24-27 September 1997*, edited by Philippe Blanchard and Arkadiusz Jadczyk, 150-55. New York: Springer, 1999.

Oriti, Daniele. "Spacetime geometry from algebra: spin foam models for non-perturbative quantum gravity." *Reports on Progress in Physics* 64 (2001): 1703-56.

Orlowski, Andrew. "Physics hoaxers discover Quantum Bogosity." *The Register*, http://www.theregister.co.uk/content/6/27894.html (June 6, 2003).

Ostriker, Jeremiah P. and Paul Steinhardt. "New Light on Dark Matter." *Science* 300 (June 20, 2003): 1909-13.

Overbye, Dennis. "A Body Scan for the Cosmos." *New York Times*, February 16, 2003, The Week in Review, 2.

------. "Are They a) Geniuses or b) Jokers? French Physicists' Cosmic Theory Creates a Big Bang of Its Own." *New York Times*, November 9, 2002, A19.

------. "String Theory: Trying to Visualize Many, Many Dimensions of Weirdness." *New York Times*, October 26, 2003, Sports, 12.

Parikh, Maulik K. "Hawking Radiation As Tunneling." hep-th/9907001 (March 2, 2001).

Park, Mary. "Lacquer on Everything." *New York Times Book Review,* June 8, 2003.

Päs, Heinrich and Thomas J. Weiler. "Absolute neutrino mass update." hep-ph/0212194 (December 13, 2002).

Paul, B. C. "Probability for primordial black holes in a higher dimensional universe." *Physical Review D* 61 (December 27, 1999): 024032 1-4.

Peat, F. David. *Superstrings and the Search for the Theory of Everything.* Lincolnwood, Ill.: Contemporary Books, 1988.

Penrose, Roger. *The Emperor's New Mind: Concerning Computers, Minds, and the Laws of Physics.* New York: Oxford University Press, 1989.

------. "Précis of *The Emperor's New Mind: Concerning computers, minds, and the laws of physics.*" *Behavioral and Brain Sciences* 13 (1990): 643-705.

------. "Quantum computation, entanglement and state reduction." *Philosophical Transactions of the Royal Society of London A* 356 (1998): 1927-39.

------. *Shadows of the Mind: A Search for the Missing Science of Consciousness.* New York: Oxford University Press, 1994.

Penrose, Roger, with Abner Shimony, Nancy Cartwright, and Stephen Hawking, Malcolm Longair, ed. *The Large, the Small and the Human Mind.* Cambridge, U. K.: Cambridge University Press, 1997.

Perez, Alejandro and Carlo Rovelli. "Spin foam model for Lorentzian general relativity." *Physical Review D* 63 (January 12, 2001): 041501 1-5.

Perez, Alejandro and Carlo Rovelli. "(3 + 1)-dimensional spin foam model of quantum gravity with spacelike and timelike components." *Physical Review D* 64 (August 24, 2001): 064002 1-12.

Pesic, Peter. *Seeing Double: Shared Identities in Physics, Philosophy, and Literature.* Cambridge, Mass.: M. I. T. Press, 2002.

Petrov, A. S. "Psychomas as Mathematical and Topological Images for Elements of Consciousness." *Radiotekhnika i Elektronika* 42 (October, 1997): 1262-65.

Piao, Yun-Song, Rong-Gen Cai, Xinmin Zhang, and Yuan-Zhong Zhang. "Assisted Tachyonic Inflation." hep-ph/0207143 (July 11, 2002).

Piao, Yun-Song, Xinmin Zhang, and Yuan-Zhong Zhang. "On Stability of the Crystal Universe Models." hep-th/0104020 (April 3, 2001).

Pitkänen, Matti. "The basic ideas of *p*-adic TGD." http://www.physics. helsinki.fi/~matpitka/padideas.html (2002).

------. "Doomsday diaries." http://www.physics.helsinki.fi/~matpitka/ diary.html.

------. "Genes, memes, qualia, and semitrance." http://www.physics. helsinki.fi/~matpitka/cbookll.html (2002).

------. "A model for biophotons." http://www.physics.helsinki.fi/ ~matpitka/articles/biophotons.pdf (2002).

------. "*p*-Adic TGD: Mathematical ideas." hep-th/9506097 (June 16, 1995).

------. "TGD inspired theory of consciousness with applications to biosystems." http://www.physics.helsinki.fi/~matpitka/ cbookl.html (2002).

------. "Topological geometrodynamics." http://www.physics.helsinki. fi/~matpitka/tgd.html (2002).

------. "Topological geometrodynamics and *p*-adic numbers." http:// www.physics.helsinki.fi/~matpitka/padtgd.html (2002).

------. "Topological geometrodynamics, Part I: General theory." *Chaos, Solitons & Fractals* 13 (2002): 1205-16.

------. "Topological geometrodynamics, Part II: Applications." *Chaos, Solitons & Fractals* 13 (2002): 1217-29.

------. "What's new in TGD and *p*-adic numbers." http://www.physics. helsinki.fi/~matpitka/newpad.html (2002).

Ponce de Leon, J. "Cosmological Models in a Kaluza-Klein Theory with Variable Rest Mass." *General Relativity and Gravitation* 20 (1988): 539-50.

Prakash, M. "Strange Pathways for Black Hole Formation." astro-ph/0009279 (September 18, 2000).

Preskill, John. "Quantum information and physics: some future directions." *Journal of Modern Optics* 47 (2000): 127-37.

Rabinowitz, Mario. "Consequences of Gravitational Tunneling Radiation." astro-ph/0302469 (March 9, 2003).

------. "Little Black Holes: Dark Matter and Ball Lightning." Redwood City, Calif.: Armor Research, lrainbow@stanford.edu (undated).

Rañada, Antonio F. "On the cosmological variation of the fine structure constant." astro-ph/0202224 (October 22, 2002).

Randall, Lisa and Raman Sundrum. "An Alternative to Compactification." *Physical Review Letters* 83 (December 6, 1999): 4690-93.

Reisenberger, Michael P. and Carlo Rovelli. "Spacetime as a Feynman diagram: the connection formulation." *Classical and Quantum Gravity* 18 (2001): 121-40.

Riordan, Michael. "Space-Time Is of the Essence." *New York Times Book Review*, August 19, 2001, 11.

Rizzo, Thomas G. "More and more indirect signals for extra dimensions at more and more colliders." *Physical Review D* 59 (May 11, 1999): 115010 1-10.

Ross, Douglas T. *Plex1: Sameness and the Need for Rigor.* Waltham, Mass.: SofTech, Inc., 1975.

Ross, Douglas T. *Plex2: Sameness and Type.* Waltham, Mass.: SofTech, Inc., 1975.

Rubakov, V. A. and M. E. Shaposhnikov. "Do We Live Inside a Domain Wall?" *Physics Letters B* 125 (May 26, 1983): 136-38.

Sagan, Carl. *Contact.* New York: Simon and Schuster, 1985.

Sakai, Osamu and Wataru Izumida. "Study on the Kondo effect in the tunneling phenomena through a quantum dot." cond-mat/0208505 (August 27, 2002).

Samal, Manoj K. "Can Science 'Explain' Consciousness?" physics/0002045 (February 24, 2000).

------. "Speculations on a Unified Theory of Matter and Mind." physics/0111035 (November 8, 2001).

Sarfatti, Jack. "Design for a Superluminal Signaling Device." *Physics Essays* 4 (1991): 315-36.

Sarid, Uri. "Tools for Tunneling." hep-ph/9804308 (July 10, 1998).

Scott, Alwyn C. "Reductionism Revisited." In *Toward a Science of Consciousness II: The Second Tucson Discussions and Debates*, edited by Stuart R. Hameroff, Alfred W. Kaszniak, and Alwyn C. Scott, 71-77. Cambridge, Mass.: M. I. T. Press, 1998.

Sen, Ashoke. "Tachyon Matter." hep-th/0203265 (March 28, 2002).

Servant, Géraldine and Tim M. P. Tait. "Is the Lightest Kaluza-Klein Particle a Viable Dark Matter Candidate?" hep-ph/0206071 (June 6, 2002).

Shapiro, James. "Confessions of a Literary Mind." *New York Times Book Review*, February 2, 2003, 18.

Shaposhnikov, Mikhail and Peter Tinyakov. "Extra dimensions as an alternative to Higgs mechanism?" *Physics Letters B* 515 (August 30, 2001): 442-46.

Shiu, Gary and S.-H. Henry Tye. "TeV scale superstring and extra dimensions." *Physical Review D* 58 (October 22, 1998): 106007 1-13.

Shiu, Gary, S.-H. Henry Tye, and Ira Wasserman. "Rolling Tachyon in

Brane World Cosmology from Superstring Field Theory." hep-th/0207119 (July 22, 2002).

Shiu, Gary and Ira Wasserman. "Cosmological Constraints on Tachyon Matter." hep-th/0205003 (June 27, 2002).

Sidharth, B. G. "The Astronomical Link Between India and the Mayans." physics/0101076 (January 18, 2001).

------. "A Brief Note on Jupiter's Magnetism." physics/9905014 (May 6, 1999).

------. "The Cosmology of Fluctuations." physics/0204073 (April 25, 2002).

------. "Duality and Cosmology." physics/0103024 (March 9, 2001).

------. "The Elusive Monopole." physics/0101014 (January 2, 2001).

------. "The Emergence of the Planck Scale." physics/0002035 (February 19, 2000).

------. "'Extended' Particles, Non Commutative Geometry and Unification." physics/0108032 (August 16, 2001).

------. "The Fractal Universe." physics/0004001 (April 1, 2000).

------. "The Fractal Universe: From the Planck to the Hubble Scale". physics/9907024 (July 17, 1999).

------. "Fuzzy, Non Commutative Space Time: A New Paradigm for a New Century." physics/0102078 (February 23, 2001).

------. "Gravitation and Electromagnetism." physics/0106051 (June 16, 2001).

------. "Gravitation and Electromagnetism II". physics/0110059 (October 20, 2001).

------. "The Holistic Universe and the Variation of the Fine Structure Constant." physics/0112094 (December 30, 2001).

------. "Instantaneous Action at a Distance in a Holistic Universe." gr-qc/9812003 (December 1, 1998).

------. "Issues and Ramifications in Quantized Fractal Space Time: An Interface with Quantum Superstrings." physics/0009084 (September 27, 2000).

------. "Monopoles and Solitons." physics/0208051 (August 13, 2002).

------. "The Nature of Space Time." physics/0204007 (April 2, 2002).

------. "Neutrino Mass and an Ever Expanding Universe (an Irreverent Perspective)." hep-ph/9811304 (November 11, 1998).

------. "The New Cosmos." physics/0211111 (November 27, 2002).

------. "A New Theory of Geomagnetism." physics/9904048 (April 23, 1999).

------. "Non-Commutative Geometry, Spin and Quarks." physics/0210057 (October 13, 2002).

------. "The Origin of Life in the Universe." physics/0205080 (May 28, 2002).

------. "Oscillator Models for the Universe and Dark Energy." physics/0302054 (February 15, 2003).

------. "Quantized Space-Time and Time's Arrow." quant-ph/9811077 (November 27, 1998).

------. "Quantum, Chaos and the Universe." quant-ph/9811045 (November 18, 1998).

------. "Quantum Mechanical Black Holes: Towards a Unification of Quantum Mechanics and General Relativity." quant-ph/9808020 (August 12, 1998).

------. "Quantum Superstrings and Quantized Fractal Space Time." physics/0010026 (October 10, 2000).

------. "A Reconciliation of Electromagnetism and Gravitation." physics/0110040 (October 14, 2001).

------. "Scale Dependent Dimensionality." physics/0004004 (April 1, 2000).

------. "The Scaled Universe." physics/9909038 (September 21, 1999).

------. "Scaled Universe II." physics/0004003 (April 1, 2000).

------. "Time-Varying Fine-Structure Constant and Quantum SuperStrings." physics/0201039 (January 19, 2002).

------. "Towards the Unification of Fundamental Interactions." physics/9905026 (May 9, 1999).

------. "The Unification of Electromagnetism and Gravitation in the Context of Quantized Fractal Space Time." physics/0007021 (July 9, 2000).

Sigurdsson, Steinn. "Testing gravity in Large Extra Dimensions using Bose-Einstein Condensates." astro-ph/0210266 (October 11, 2002).

Siler, Todd. *Breaking the Mind Barrier: The Artscience of Neurocosmology.* New York: Simon and Schuster, 1990.

Sirag, Saul-Paul. "A Mathematical Strategy for a Theory of Consciousness." In *Toward a Science of Consciousness: The First Tucson Discussions and Debates*, edited by Stuart R. Hameroff, Alfred W. Kaszniak, and Alwyn C. Scott, 579-88. Cambridge, Mass.: M. I. T. Press, 1996.

Smilga, A. V. "Quantum gravity as Escher's dragon". hep-th/0212033 (December 3, 2002).

Smith, David Alexander. *In the Cube.* New York: Tom Dohrety Associates, 1993.

Smolin, Lee. *Three Roads to Quantum Gravity.* New York: Basic Books, 2001.

Solodukhin, Sergey N. "Black Hole Production Via Quantum Tunneling." hep-th/0212001 (November 30, 2002).

Spaans, Marco. "Observational Tests to Discern the Topology of

Planckian Space-Time." hep-th/9811142 (November 18, 1998).

------. "On the Topological Nature of Fundamental Interactions." gr-qc/9901025 (October 11, 2000).

------. "Topological Dynamics and Grand Unified Theory." gr-qc/9704036 (April 14, 1997).

------. "A Topological Extension of General Relativity." gr-qc/9612027 (December 11, 1996).

------. "A Topological Formulation of the Standard Model." gr-qc/9711048 (November 15, 1997).

------. "Towards a Topological Formulation of Fundamental Interactions." hep-th/9806172 (June 19, 1998).

Sridhar, K. "Large Extra Dimensions at Linear Colliders." *International Journal of Modern Physics A* 15 (2000): 2397-2403.

Stapp, Henry P. "Attention, Intention, and Will in Quantum Physics." quant-ph/9905054 (May 17, 1999).

------. "Quantum Ontology and Mind-Matter Synthesis." In *Quantum Future: From Volta and Como to the Present and Beyond; Proceedings of the Xth Max Born Symposium Held in Przesieka, Poland, 24-27 September 1997*, edited by Philippe Blanchard and Arkadiusz Jadczyk, 156-203. New York: Springer, 1999.

------. "Quantum Theory and the Role of Mind in Nature." *Foundations of Physics* 31 (October 2001): 1465-99.

------. "The 18-Fold Way." *Foundations of Physics* 32 (February 2002): 255-66.

Starkman, Glenn D., Dejan Stojkovic, and Mark Trodden. "Homogeneity, Flatness, and 'Large' Extra Dimensions." *Physical Review Letters* 87 (December 3, 2001): 231303 1-4.

Stecker, Floyd W. "Implications of Ultrahigh Energy Air Showers for Physics and Astrophysics." astro-ph/0207629 (August 12, 2002).

Stelle, K. S. "The unification of quantum gravity." *Nuclear Physics B (Proceedings Supplement)* 88 (2000): 3-9.

Stephens, G. J. and B. L. Hu. "Notes on Black Hole Phase Transition." *International Journal of Theoretical Physics* 40 (December, 2001): 2183-2200.

Stienstra, Jan. "Ordinary Calabi-Yau-3 Crystals." math.AG/0212061 (December 18, 2002).

Stoica, Horace, S.-H. Henry Tye, and Ira Wasserman. "Cosmology in the Randall-Sundrum Brane World Scenario." hep-th/0004126 (April 17, 2000).

Strehle, Susan. *Fiction in the Quantum Universe.* Chapel Hill: University of North Carolina Press, 1992.

Suplee, Curt. *Physics in the 20th Century*. New York: Harry N. Abrams, 1999.

Tegmark, Max and Alexander Vilenkin. "Anthropic predictions for neutrino masses." astro-ph/0304536 (April 29, 2003).

Terashima, Hiroaki. "Entanglement entropy of the black hole horizon." *Physical Review D* 61 (April 25, 2000): 104016 1-11.

Ter-Kazarian, G. T. "Introduction to the Theory of Goyaks: Operator Manifold Approach to Geometry and Particle Physics. hep-th/9510110 (October 16, 1995).

Tkachenko, Alexei V. "Electrostatic effects in DNA stretching." cond-mat/0303076 (March 5, 2003).

Tomsovic, S. "Tunneling and Chaos." nlin.CD/0008031 (June 12, 2001).

Traub, James. "Scary." *New York Times Magazine*, May 9, 2004, 13-14.

Trofimenko, A. P. and V. S. Gurin. "White and Grey Holes in Higher-Dimensional Representation of Extended Space-Time Manifolds of General Relativity." *Annalen der Physik* 48 (1991): 295-303.

Tsubota, Makoto and Kenichi Kasamatsu. "Josephson Current Flowing in Cyclically Coupled Bose-Einstein Condensates." cond-mat/9911389 (May 28, 2000).

Turner, Michael S. "Dark Matter and Dark Energy: The Critical Questions." astro-ph/0207297 (July 14, 2002).

Ubriaco, M. R. "Fermions on the Field of p-Adic Numbers." *Physical Review D* 41 (April 15, 1990): 2631-33.

Ulmschneider, Peter. *Intelligent Life in the Universe: From Common Origins to the Future of Humanity*. New York: Springer-Verlag, 2003.

Veale, Scott. "New & Noteworthy Paperbacks." *New York Times Book Review*, August 18, 2002, 20.

Velmans, Max. "Goodbye to Reductionism." In *Toward a Science of Consciousness II: The Second Tucson Discussions and Debates*, edited by Stuart R. Hameroff, Alfred W. Kaszniak, and Alwyn C. Scott, 45-52. Cambridge, Mass.: M. I. T. Press, 1998.

Vertogradova, E. G. and Yu. S. Grishkan. "A Self-Consistent Model for an Isotropic Universe Evolving under the Influence of Quantum Gravitational Effects." *Astronomy Reports* 44 (2000): 142-49.

Vilenkin, Alexander. "The quantum cosmology debate." gr-qc/9812027 (December 7, 1998).

Vladimirov, V. S. and Ya. I. Volovich. "On the Nonlinear Dynamical Equation in the p-Adic String Theory." math-ph/0306018 (June 5, 2003).

Walker, Gabrielle. "Here Comes Hypertime." *New Scientist* 156 (November 1, 1997): 40-41.

Wallraff, A., A. Lukashenko, C. Coqui, T. Duty, and A. V. Ustinov. "High resolution measurements of the switching current in a Josephson tunnel junction: Thermal activation and macroscopic quantum tunneling." cond-mat/0204527 (August 29, 2002).

Waltham, Chris. "Neutrino Oscillations for Dummies." physics/0303116 (March 28, 2003).

Wang, Bin, Elcio Abdalla, and Ru-Keng Su. "Dynamics and holographic discreteness of tachyonic inflation." hep-th/0208023 (August 3, 2002).

Webb, Stephen. *If the Universe Is Teeming with Aliens . . . Where Is Everybody?: Fifty Solutions to the Fermi Paradox and the Problem of Extraterrestrial Life.* New York: Copernicus Books, 2002.

Wesson, Paul S. and Hongya Liu. "Shell-like solutions in Kaluza-Klein theory." *Physics Letters B* 432 (July 30, 1998): 266-270.

Wigmans, Richard. "Are Massive Neutrinos Responsible for the Accelerated Expansion of the Universe?" astro-ph/0210272 (October 26, 2002).

Wigner, Eugene P. "The Unreasonable Effectiveness of Mathematics in the Natural Sciences." *Communications on Pure and Applied Mathematics* XIII (1960): 1-14.

Wolf, Fred Alan. "On the Quantum Mechanics of Dreams and the Emergence of Self-Awareness." In *Toward a Science of Consciousness: The First Tucson Discussions and Debates*, edited by Stuart R. Hameroff, Alfred W. Kasizniak, and Alwyn C. Scott, 451-67. Cambridge, Mass.: M. I. T. Press, 1996.

------. "On the Quantum Physical Theory of Subjective Antedating." *Journal of Theoretical Biology* 136 (1989): 13-19.

------. *Parallel Universes: The Search for Other Worlds.* New York: Simon and Schuster, 1988.

------. *Star Wave: Mind, Consciousness, and Quantum Physics.* New York: Macmillan, 1984.

------. *Taking the Quantum Leap: The New Physics for Nonscientists.* New York: Harper and Row, 1989.

Wolfram, Stephen. *A New Kind of Science.* Champaign, Ill.: Wolfram Media, 2002.

Yasue, Kunio. "Quantum Monadology." In *Toward a Science of Consciousness III: The Third Tucson Discussions and Debates*, edited by Stuart R. Hameroff, Alfred W. Kasizniak, and David J. Chalmers, 317-327. Cambridge, Mass.: M. I. T. Press, 1999.

Youm, Donam. "A Varying-alpha Brane World Cosmology." hep-th/0108237 (September 5, 2001).

Zaman III, L. F. "Nature's Psychogenic Forces: Localized Quantum Consciousness." *Journal of Mind and Behavior* 23 (Autumn, 2002): 351-74.

Zapata, Ivar, Fernando Sols, and Anthony J. Leggett. "Josephson effect between trapped Bose-Einstein condensates." cond-mat/9707143 (July 14, 1997).

Zhou, Fei. "Hidden Topological Order in ^{23}Na ($F = 1$) Bose-Einstein Condensates." cond-mat/0107263 (July 13, 2001).

Zlotowitz, Rabbi Meir and Rabbi Nosson Scherman. *Bereshis/Genesis: A New Translation with a Commentary Anthologized from Talmudic, Midrashic and Rabbinic Sources.* New York: Mesorah Publications, Ltd., 1977.

Zohar, Danah. "Consciousness and Bose-Einstein Condensates." In *Toward a Science of Consciousness: The First Tucson Discussions and Debates*, edited by Stuart R. Hameroff, Alfred W. Kaszniak, and Alwyn C. Scott, 439-50. Cambridge, Mass.: M. I. T. Press, 1996.

------. "A Quantum Mechanical Model of Consciousness and the Emergence of 'I.'" *Minds and Machines* 5 (1995): 597-607.

------. *The Quantum Self: Human Nature and Consciousness Defined by the New Physics.* New York: William Morrow, 1990.

Zukav, Gary. *The Dancing Wu Li Masters: An Overview of the New Physics.* New York: William Morrow, 1979.

Endnotes

[1] The discussion of Solvay and the Solvay Conferences is based on Mehra, *The Solvay Conferences on Physics*, 1-3.

[2] Blanchard and Jadczyk, *Quantum Future*, v.

[3] Stapp, "Quantum Ontology and Mind-Matter Synthesis," 156.

[4] The "twin paradox" is frequently referred to in discussions of special relativity. I first encountered this in French, *Special Relativity*.

[5] Mauceli, Geng, Hamilton, Johnson, Merkowitz, Morse, Price, and Solomonson, "The Allegro gravitational wave detector."

[6] Cerdonio, "The search for gravitational waves."

[7] Hepp, "Toward the Demolition of a Computational Quantum Brain."

[8] Popping the qwiff" is Wolf's riff on "quantum wave function" in *Taking the Quantum Leap*.

[9] Lyre, "Against Measurement?"

[10] Kiefer and Joos, "Decoherence."

[11] Audretsch and Mensky, "Continuous Fuzzy Measurements and Visualization of a Quantum Transition."

[12] Omnés, "On Interpretation and Decoherence."

[13] Carmichael, "Quantum Jumps Revisited," 16.

[14] Penrose's full discussion of physics' "superb" and other theories, which I summarize in this chapter, appears on pages 151-224 of *The Emperor's New Mind*.

[15] Davies and Brown, *Superstrings*, 133-34.

[16] Kaku, *Hyperspace*, 171.

[17] Davies and Brown, *Superstrings*, 191.

18 From Overbye, "String Theory," quoting the Public Broadcasting System series "The Elegant Universe," which in turn quoted Edward Witten.

19 Peat, *Superstrings and the Search for the Theory of Everything*, 319-21.

20 *Flatland: A Romance of Many Dimensions*, originally published in 1884, has been republished a number of times.

21 This chapter's assignment of physicists to various schools of quantum reality is based on Herbert, *Quantum Reality.*

22 Herbert, *Quantum Reality*, 248-50.

23 This chapter has drawn from a number of essays on reductionism and its role in science: Castellani, "Reductionism, Emergence, and Effective Field Theories"; Grössing, *Quantum Cybernetics*, 1-8; Marshall and Zohar, *Who's Afraid of Schrödinger's Cat?*, 307-8; and Scott, "Reductionism Revisited."

24 Norris, *Quantum Theory and the Flight from Realism*, 1-5.

25 Ibid., 5.

26 Velmans, "Goodbye to Reductionism."

27 Hazen, "The Great Unanswered Questions."

28 Hawking and Penrose, *The Nature of Space and Time*, 124.

29 Kragh, *Quantum Generations*, 242, quoting Alexander Vucinich's 1980 article in *Isis*, "Soviet physicists and philosophers in the 1930s: Dynamics of a conflict."

30 Kragh, *Quantum Generations*, 212.

31 Ibid., 210, noting Bohr's comments at the 1938 International Congress of Anthropolgical and Ethnological Sciences.

32 Ibid., 212, quoting from Max Jammer's 1974 *The Philosophy of Quantum Mechanics: The Interpretation of Quantum Mechanics in Historical Perspective.*

33 Penrose, "Précis of *The Emperor's New Mind.*"

34 Penrose, *The Emperor's New Mind*, 222.

35 The discussion of Penrose's speculations on the nature of consciousness is my summary of pages 405-449 of *The Emperor's New Mind.*

36 Penrose, *The Emperor's New Mind*, 423.

37 The discussion of the one-graviton level draws from Penrose, *The Emperor's New Mind*, 367-372.

38 Hameroff, "Did Consciousness Cause the Cambrian Evolutionary Explosion?"

39 Penrose, *The Emperor's New Mind*, 447.

40 Penrose, "Précis of *The Emperor's New Mind*," 643-705.

41 The discussion of Wolf's *Taking the Quantum Leap* is based on pages 228-45.

42 Bass, "A Quantum Mechanical Mind-Body Interaction."

43 Wolf, "On the Quantum Physical Theory of Subjective Antedating."

[44] Eccles, "Do Mental Events Cause Neural Events?"

[45] Eccles, *Evolution of the Brain.*

[46] The discussion of dendrons and psychons is my summary of pages 186-92 of Eccles' *Evolution of the Brain*, and pages 549-72 of *The Principles of Design and Operation of the Brain*, which comprises the proceedings of a study week organized by the Pontifical Academy of Sciences, edited by Eccles and Creutzfeldt.

[47] Liu, "Neurons, Psychons, and Emotion."

[48] Kitcher, "The Mind Mystery."

[49] Donald, "Quantum Theory and the Brain."

[50] Mould, "Consciousness and Endogenous State Reduction," "Consciousness and Quantum Mechanics," and "Quantum Consciousness."

[51] Stapp, "Attention, Intention, and Will in Quantum Physics" and "Quantum Theory and the Role of Mind in Nature."

[52] Stapp, "Quantum Theory and the Role of Mind in Nature," 1490.

[53] Stapp, "The 18-Fold Way."

[54] Kafatos and Nadeau, *The Conscious Universe*, 139.

[55] Marshall and Zohar discuss holism and emergence in *Who's Afraid of Schrödinger's Cat?*, 184-87 and 137-39, respectively.

[56] Ali and Zimmer, "Beyond Substance and Process, " 585.

[57] Discussed by Wolf in *Star Wave*, 267-325.

[58] The discussion of Explicate Order and Implicate Order is based on Bohm, *Wholeness and the Implicate Order*, 111-57 and 179-213, and Marshall and Zohar, *Who's Afraid of Schrödinger's Cat?*, 197-99.

[59] Zohar, "A Quantum Mechanical Model of Consciousness."

[60] Zohar, *The Quantum Self.*

[61] Marshall, "Some phenomenological implications of a quantum model of consciousness."

[62] Zohar, *The Quantum Self*, 52.

[63] Zohar, "A Quantum Mechanical Model of Consciousness," 597.

[64] Kafatos and Nadeau, *The Conscious Universe*, 10-11.

[65] Poor Lord Kelvin. He's also quoted as predicting, in 1897, that "radio has no future" (sidebar to Traub, "Scary," 13).

[66] Smolin, *Three Roads to Quantum Gravity*, 131.

[67] Ibid., 125-45.

[68] Grosse and Schlesinger, "Spin foam models for M-theory."

[69] Mikovi , "Spin foam models of matter coupled to gravity"; Oriti, "Spacetime geometry from algebra"; Perez and Rovelli, "Spin foam model for Lorentzian general relativity"; Reisenberger and Rovelli, "Spacetime as a Feynman diagram."

[70] Smolin, *Three Roads to Quantum Gravity*, 178.

[71] Riordan, "Space-Time Is of the Essence."

[72] Gribbin, *Q Is for Quantum*, 219.

[73] Ibid., 317-20.

[74] Giudice, Rattazzi, and Wells, "Quantum gravity and extra dimensions at high-energy colliders."

[75] Emparan, Masip, and Rattazzi, "Cosmic rays as probes of large extra dimensions and TeV gravity."

[76] Vertogradova and Grishkan, "A Self-Consistent Model for an Isotropic Universe."

[77] Carlip, "Quantum gravity," 925.

[78] Kaku, *Hyperspace*, 21.

[79] Kaku's historical context for extra dimensions is at *Hyperspace*, 20-23.

[80] The history of early Kaluza-Klein theories is based on Gribbin, *Q Is for Quantum*, 196-97.

[81] Early models of compactified extra dimensions are discussed in Antoniadis, "A possible new dimension at a few TeV," and Rubakov and Shaposhnikov, "Do We Live Inside a Domain Wall?" Extra dimensions with size up to a millimeter are discussed in Arkani-Hamed, Dimopoulos, and Dvali, "The hierarchy problem and new dimensions at a millimeter" and "Phenomenology, Astrophysics and Cosmology"; Dvali, "Submillimeter Extra Dimensions"; and elsewhere.

[82] Models with infinite-size extra dimensions are discussed in Arkani-Hamed, Dimopoulos, Dvali, and Kaloper, "Infinitely Large New Dimensions"; Corradini, Iglesias, Kakushadze, and Langfelder, "Gravity on a 3-brane"; Dvali, Gabadadze, and Porrati, "Metastable gravitons and infinite volume extra dimensions"; Gregory, Rubakov, and Sibiryakov, "Opening up Extra Dimensions"; and Randall and Sundrum, "An Alternative to Compactification."

[83] The discussion of particle accelerators draws from Gribbin, *Q Is for Quantum*, 11-12.

[84] Extra dimensions are widely discussed. Useful sources on finding extra dimensions include Arkani-Hamed, Dimopoulos, and Dvali, "The hierarchy problem and new dimensions at a millimeter," and Rizzo, "More and more indirect signals for extra dimensions."

[85] Grand unification through universal extra dimensions is discussed in Appelquist, Cheng, and Dobrescu, "Bounds on universal extra dimensions"; Dienes, Dudas, and Gherghetta, "Extra spacetime dimensions and unification," "TeV-Scale GUTs," and "Grand unification at intermediate mass scales"; and Shiu and Tye, "TeV scale superstring and extra dimensions."

[86] Sridhar, "Large Extra Dimensions at Linear Colliders."

[87] Dienes, Dudas, and Gherghetta, "Extra spacetime dimensions and unification," "TeV-Scale GUTs," and "Grand unification at intermediate mass scales."

[88] Dienes, "Shape versus Volume."

[89] Piao, Zhang, and Zhang, "On Stability of the Crystal Universe Models."

[90] Arkani-Hamed, Dimopoulos, and Dvali, "Phenomenology, Astrophysics and Cosmology."

[91] Castro, "Is quantum space-time infinite dimensional."

[92] Levin, "Topology and the Cosmic Microwave Background."

[93] Inflation is a widely discussed theory in journals of physics. Alan Guth was the original developer of the theory of inflation, which he published in a 1981 *Physical Review* article titled "Inflationary universe: A possible solution to the horizon and flatness problems." Guth's *The Inflationary Universe: The Quest for a New Theory of Cosmic Origins* was a particularly useful source for this chapter. Note in particular that the specific numerical quantities associated with inflation—including those related to the timing and pace of inflation—are from *The Inflationary Universe*; but, as Guth points out, these quantities are not at this time determined with certainty. The text's *illustrative* values are from Guth, *The Inflationary Universe*, 181-86.

[94] Guth, *The Inflationary Universe*, 177-78.

[95] Khoury, Ovrut, Steinhardt, and Turok, "Ekpyrotic universe."

[96] Easther, Greene, Kinney, and Shiu, "Inflation as a Probe of Short Distance Physics."

[97] Stoica, Tye, and Wasserman, "Cosmology in the Randall-Sundrum Brane World Scenario."

[98] Guth, *The Inflationary Universe*, 281.

[99] Smolin, *Three Roads to Quantum Gravity*, 26-32.

[100] Krauss and Starkman, "Life, the Universe, and Nothing."

[101] Gudmundsson and Björnsson, "Dark Energy and the Observable Universe."

[102] Guendelman and Owen, "Universe Creation, Entropy, and Extra Dimensions."

[103] irkovi and Bostrom, "Cosmological Constant and the Final Anthropic Hypothesis."

[104] Cornish and Spergel, "A small universe after all?"

[105] Inoue and Sugiyama, "How large is our universe?"

[106] Marshall and Zohar, Who's Afraid of Schrödinger's Cat?, 203-5.

[107] Abbott, Barr, and Ellis, "Kaluza-Klein cosmologies and inflation."

[108] Levin, "Inflation from extra dimensions."

[109] Camp and Safko, "Quantization Conditions in Curved Spacetime." This is not actually an extradimensional model of inflation, but one which derives from the effects of curved spacetime on how quantum physics' uncertainty principle operates.

[110] Arkani-Hamed, Dimopoulos, Kaloper, and March-Russell, "Early

Inflation and Cosmology"; Burgess, Majumdar, Nolte, Quevedo, Rajesh, and Zhang, "The Inflationary Brane-Antibrane Universe"; Flanagan, Tye, and Wasserman, "Cosmology of the brane world"; Jones, Stoica, and Tye, "Brane Interaction as the Origin of Inflation"; Mongan, "(N + 1)-Dimensional Quantum Mechanical Model."

[111] Cheng, Feng, and Matchev, "Kaluza-Klein Dark Matter."
[112] Cheng, Feng, and Matchev, "Kaluza-Klein Dark Matter"; Demir, "Stable Q-balls from extra dimensions"; Servant and Tait, "Is the Lightest Kaluza-Klein Particle a Viable Dark Matter Candidate?"
[113] Turner, "Dark Matter and Dark Energy."
[114] Gardner, "Primordial inflation and present-day cosmological constant."
[115] Deffayet and Dvali, "Accelerated universe from gravity leaking to extra dimensions."
[116] Turner, "Dark Matter and Dark Energy."
[117] Original data and categorization from Turner, "Dark Matter and Dark Energy,"updated for the latest astronomical observations, reported in Ostriker and Steinhardt, "New Light on Dark Matter" and Overbye, "A Body Scan for the Cosmos."
[118] Durrer, "Frontiers of the Universe."
[119] Fairbairn and Tytgat, "Inflation from a Tachyon Fluid?"; Gibbons, "Cosmological Evolution of the Rolling Tachyon"; Li, Liu, and Hao, "On the tachyon inflation"; Mazumdar, Panda, and Pérez-Lorenzana, "Assisted inflation via tachyon condensation"; Sen, "Tachyon Matter"; Shiu, Tye, and Wasserman, "Rolling Tachyon in Brane World Cosmology"; Wang, Abdalla, and Su, "Dynamics and holographic discreteness of tachyonic inflation."
[120] Ni, "Evidence for Neutrino Being Likely a Superluminal Particle."
[121] Shiu and Wasserman, "Cosmological Constraints on Tachyon Matter."
[122] Kofman and Linde, "Problems with Tachyon Inflation."
[123] Piao, Cai, Zhang, and Zhang, "Assisted Tachyonic Inflation."
[124] Barrow and Kodama, "All Universes Great and Small"; Starkman, Stojkovic, and Trodden, "Homogeneity, Flatness, and 'Large' Extra Dimensions."
[125] Khoury, Ovrut, Steinhardt, and Turok, "Ekpyrotic universe."
[126] Albrecht and Magueijo, "Time varying speed of light as a solution to cosmological puzzles."
[127] Levin and Freese, "Possible solution to the horizon problem."
[128] Durrer and Laukenmann, "The oscillating universe."
[129] Penrose, The Emperor's New Mind, 347.
[130] Kent, "Night Thoughts of a Quantum Physicist."
[131] Camp and Safko, "Quantization Conditions in Curved Spacetime."

[132] Carugno, Litterio, Occhionero, and Pollifrone, "Inflation in multidimensional quantum cosmology."

[133] Li, Liu, and Hao, "On the tachyon inflation."

[134] Hollands and Wald, "An Alternative to Inflation."

[135] Smolin, *Three Roads to Quantum Gravity*, 205.

[136] Durrer, "Frontiers of the Universe."

[137] The discussion of the EPR paradox draws from Gribbin, *Q Is for Quantum*, 126-27, and Herbert, *Quantum Reality*, 201-4.

[138] The discussion of the Aspect experiment draws from Gribbin, *Q Is for Quantum*, 22-24; Herbert, *Quantum Reality*, 219-27; and Penrose, *The Emperor's New Mind*, 286-87.

[139] Gribbin, *Q Is for Quantum*, 109, quoting Richard Feynman.

[140] The discussion of nonlocality and the double-slit experiment draws from Gribbin, *Q Is for Quantum*, 109-13 and 258.

[141] Penrose, "Quantum computation, entanglement and state reduction."

[142] Marshall and Zohar, *Who's Afraid of Schrödinger's Cat*, 176.

[143] Preskill, "Quantum information and physics," 128.

[144] Ibid., 131.

[145] Sarfatti, "Design for a Superluminal Signaling Device."

[146] Lewenstein, Bruß, Cirac, Kraus, Kus, Samsonowicz, Sanpera, and Tarrach, "Separability and distillability in composite quantum systems," 2482.

[147] Penrose, *The Emperor's New Mind*, 306-9.

[148] Kaku, *Hyperspace*, 305-6.

[149] Greene, *The Elegant Universe*, 333-40.

[150] Barbour, *The End of Time*, 25.

[151] Ferris, *The Mind's Sky*, 211.

[152] Lyre, "Against Measurement?"

[153] Smolin, *Three Roads to Quantum Gravity*.

[154] Gribbin, *Q Is for Quantum*, 47.

[155] Rabinowitz, "Little Black Holes."

[156] The discussion of aging star processes and black hole formation draws from Gribbin, *Q Is for Quantum*, 45-48, and Kaku, *Hyperspace*, 217-31.

[157] A discussion of wormholes is included within González-Díaz, "Observable effects from spacetime tunneling."

[158] Smolin, *Three Roads to Quantum Gravity*.

[159] Wolf, *Parallel Universes*.

[160] Hawking, *A Brief History of Time*.

[161] Feng and Shapere, "Black Hole Production by Cosmic Rays."

[162] Paul, "Probability for primordial black holes in a higher dimensional universe."

[163] Argyres, Dimopoulos, and March-Russell, "Black holes and sub-millimeter dimensions."

[164] Berezin, "Quantum Black Hole."

[165] Maia and Silveira, "Window to extra dimensions near a black hole."

[166] Dzhunushaliev, "Domain with Noncompactified Extra Dimensions."

[167] Emparan, Horowitz, and Myers, "Black Holes Radiate Mainly on the Brane"; Giddings and Thomas, "High energy colliders as black hole factories."

[168] Chamblin, Csáki, Erlich, and Hollowood, "Black diamonds at brane junctions."

[169] Casadio and Harms, "Black hole evaporation and large extra dimensions" and "Black hole evaporation and compact extra dimensions."

[170] Hosoya, Carlini, and Shimomura, "Generalized second law of black hole thermodynamics."

[171] Terashima, "Entanglement entropy of the black hole horizon."

[172] Gurin and Trofimenko, "Higher-dimensional white holes" and "White Holes in Kaluza-Klein Theory"; Stephens and Hu, "Notes on Black Hole Phase Transition."

[173] The discussion of Hawking's work is based on Hawking, *A Brief History of Time*, *The Theory of Everything*, and *The Universe in a Nutshell*.

[174] Kaku, *Hyperspace*, 252-69.

[175] The discussion of alpha decay and nuclear fusion as tunneling phenomena draws from Gribbin, *Q Is for Quantum*, 15, 150-51, and 414.

[176] The discussion of scanning tunneling microscopes draws from Suplee, *Physics in the 20th Century*, 32-35, 81, 86-87, and 114.

[177] Loss and DiVincenzo, "Quantum Computation with Quantum Dots"; Sakai and Izumida, "Study on the Kondo Effect."

[178] Vilenkin, "The quantum cosmology debate."

[179] Sarid, "Tools for Tunneling."

[180] Solodukhin, "Black Hole Production Via Quantum Tunneling."

[181] Parikh and Wilczek, "Hawking Radiation as Tunneling."

[182] Rabinowitz, "Consequences of Gravitational Tunneling Radiaition."

[183] De Vries and Schmidt-Kaler, "The black hole tunnel phenomenon."

[184] De Carvalho and Cavalcanti, "Tunnelings as catastrophe"; Horwitz, Levitan, and Ashkenazy, "Complexity, Tunneling and Geometrical Symmetry"; Keshavamurthy, "Dynamical tunneling in molecules"; Tomsovic, "Tunneling and Chaos."

[185] Klipa and Bosanac, "Understanding quantum effects from classical principles"; Olavo, "Quantum Mechanics as a Classical Theory XIII."

[186] Chiao, "Tunneling Times and Superluminality"; Garcia-Calderón, "Early times in tunneling."

[187] Davidson, "Tachyons, Quanta and Chaos."

[188] Horwitz, "Time and Entropy from Semi-classical Tunneling."

[189] Nimtz, "Superluminal Tunneling Devices."

[190] The discussion of Brian Josephson and of Josephson junctions draws from Gribbin, *Q Is for Quantum* , 195-96, and Suplee, *Physics in the 20th Century*, 81.

[191] Gribbin, *Q Is for Quantum*, 195.

[192] Abanov and Wiegmann, "Tunneling in the topological mechanism of superconductivity."

[193] Wallraff, Lukashenko, Coqui, Duty, and Ustinov, "High resolution measurements of the switching current in a Josephson tunnel junction."

[194] Averin, "Quantum computing and quantum measurement with mesoscopic Josephson junctions."

[195] The history of Bose-Einstein condensates draws from Gribbin, *Q Is for Quantum*, 57-58.

[196] Ketterle, Durfee, and Stamper-Kurn, "Making, probing and understanding Bose-Einstein condensates."

[197] Moore and Meystre, "Generating entangled atom-photon pairs from Bose-Einstein condensates."

[198] De Oliveira, "Teleportation of a Bose-Einstein condensate."

[199] Burioni, Cassi, Meccoli, Rasetti, Regina, Sodano, and Vezzani, "Bose-Einstein condensation in inhomogeneous Josephson arrays"; Chen and Zhang, "Possible realization of Josephson charge qubits in two-coupled Bose-Einstein condensates"; Dziarmaga, Smerzi, Zurek, and Bishop, "Dynamics of Quantum Phase Transition in an Array of Josephson Junctions"; Giovanazzi, Smerzi, and Fantoni, "Josephson effects in dilute Bose-Einstein condensates"; Tsubota and Kasamatsu, "Josephson Current Flowing in Cyclically Coupled Bose-Einstein Condensates"; Zapata, Sols, and Leggett, "Josephson effect between trapped Bose-Einstein condensates."

[200] Dymnikova and Khlopov, "Decay of cosmological constant as Bose condensate evaporation."

[201] Mazur and Mottola, "Gravitational Condensate Stars."

[202] Barceló, Liberati, and Viser, "Analog gravity from Bose-Einstein condensates."

[203] Zhou, "Hidden Topological Order in ^{23}Na Bose-Einstein Condensates."

[204] Barceló and Campo, "Braneworld physics from the analog-gravity perspective."

[205] Sigurdsson, "Testing gravity in Large Extra Dimensions using Bose-Einstein Condensates."

[206] Chiao, "Conceptual tensions between quantum mechanics and general relativity."

[207] Marshall and Zohar, *Who's Afraid of Schrödinger's Cat?*, 76.

[208] Faisal, "Quantum Chaos, Algorithmic Paradigm and Irreversibility," 48.

[209] This section draws from Suplee, *Physics in the 20th Century*, 152-79.

[210] Castro, "Is quantum space-time infinite dimensional."

[211] Argyris, Ciubotariu, and Weingaertner, "Fractal space signatures in quantum physics and cosmology."

[212] Samal, "Speculations on a Unified Theory of Matter and Mind."

[213] The preliminary paragraphs on neutrinos draw from Greene, *The Elegant Universe*, 7-12, and Kaku, *Hyperspace*, 125, 128, and 187-88.

[214] Päs and Weiler, "Absolute neutrino mass update."

[215] Haxton, "Nuclear Problems in Astrophysics."

[216] Auriemma, "High energy neutrinos."

[217] Dolgov, "Neutrinos in cosmology."

[218] Wigmans, "Are Massive Neutrinos Responsible for the Accelerated Expansion of the Universe?"

[219] Prakash, "Strange Pathways for Black Hole Formation."

[220] Ellis, "Particle Candidates for Dark Matter."

[221] Caban, Rembieli ski, Smoli ski, and Walczak, "Oscillations do not distinguish between massive and tachyonic neutrinos"; Ni, "Evidence for Neutrino Being Likely a Superluminal Particle."

[222] Ammosov and Volkov, "Can Neutrinos Probe Extra Dimensions?"; Dvali and Smirnov, "Probing Large Extra Dimensions with Neutrinos."

[223] Caldwell, Mohapatra, and Yellin, "Neutrinos, Large Extra Dimensions and Solar Neutrino Puzzle"; Dighe and Joshipura, "Neutrino anomalies and large extra dimensions"; Ling, "A minimal model with large extra dimensions to fit the neutrino data."

[224] Miramonti and Reseghetti, "Solar neutrino physics."

[225] Olinto, "Messengers of the Extreme Universe."

[226] Waltham, "Neutrino Oscillations for Dummies."

[227] Bhattacharya and Pal, "Neutrinos and magnetic fields."

[228] Tegmark and Vilenkin, "Anthropic predictions for neutrino masses."

[229] Asimov, *The Neutrino: Ghost Particle of the Atom*, xiii.

[230] For example, ninety-eight researchers from twelve institutions in the KamLAND Collaboration co-authored Eguchi et al., "First Results from KamLAND: Evidence for Reactor Antineutrino Disappearance," and 122 and 123 researchers (respectively) from twenty-nine institutions in the Super-Kamiokande Collaboration co-authored Malek et al., "Search for Supernova Relic Neutrinos at Super-Kamiokande" and Gando et al., "Search for *ve* from the Sun at Super-Kamiokande-I."

[231] Wigner, "The Unreasonable Effectiveness of Mathematics," 2.

[232] Ibid., 6.

233 Ibid., 7.

234 Ibid., 8.

235 Ibid., 14.

236 There are numerous formulations of Maxwell's equations. These formulations, of both the set of four equations and their single-equation compression, are from Kaku, *Hyperspace*, 342. The correct transcription of these equations requires the four deltas to be written upside down, as the operator del, not as the Greek letter delta. Also, the multiplication in the second and fourth of the four equations is cross product multiplication.

237 Barbour, *The End of Time*, 16.

238 Ibid., 177.

239 Ibid., 194.

240 Strong determinism and many-worlds views are discussed at Penrose, *The Emperor's New Mind*, 432.

241 Barbour, *The End of Time*, 167.

242 Demir, "Stable Q-balls from extra dimensions."

243 This work of Harvard physicist Cumrun Vafa is summarized in Walker, "Here Comes Hypertime."

244 This discussion of inertial and gravitational mass draws from Gribbin, *Q Is for Quantum*, 165, 184, and 217-20.

245 Lederman and Teresi, *The God Particle*, 341.

246 Ibid., 375.

247 Ibid., 375.

248 The discussion of "marble" and "wood" draws from Kaku, *Hyperspace*, 98-99.

249 Ponce de Leon, "Cosmological Models in a Kaluza-Klein Theory"; Wesson and Liu, "Shell-like soluitons in Kaluza-Klein Theory."

250 Mak and Harko, "Cosmological particle production in five-dimensional Kaluza-Klein theory."

251 Shaposhnikov and Tinyakov, "Extra dimensions as an alternative to Higgs mechanism?"

252 Frampton, Hung, and Sher, "Quarks and Leptons Beyond the Third Generation."

253 Pitkänen, "The basic ideas of *p*-adic TGD."

254 Caban, Rembieli ski, Smoli ski, and Walczak, "Oscillations do not distinguish between massive and tachyonic neutrinos."

255 Kirn, "Winging It."

256 Siler, *Breaking the Mind Barrier*, 189.

257 Ibid., 130.

258 Ibid., 57.

259 Ibid., 92.

260 Guth, *The Inflationary Universe*, xi.

[261] The reference to ethnomathematics is based on Olin, "Ethnomathematics."

[262] Muschamp, "A See-Through Library of Shifting Shapes and Colors." Muschamp is quoting historian Stephen Kern with the phrase "the culture of time and space."

[263] Karlin, "A Sculptor Works Up an Exposé of the Stars' Secrets."

[264] Jonas, "Science Fiction."

[265] Mendelsohn, "A Dance to the Music of Time." Mendelsohn's comments are in review of *The Time of Our Singing*, by Richard Powers.

[266] Shapiro, "Confessions of a Literary Mind." Shapiro's comments are in review of *The Child That Books Built: A Life in Reading*, by Francis Spufford.

[267] Veale, "New & Noteworthy Paperbacks."

[268] Ferguson, "Books in Brief." Ferguson's comments are in review of *The Here and Now*, by Gregg Easterbrook.

[269] The "conventions of quantum physics" are deployed in *The Strength of the Sun*, by Catherine Chidgey, and are discussed in an unattributed review appearing in the May 26, 2002, *New York Times Book Review*.

[270] Strehle, *Fiction in the Quantum Universe*, x.

[271] Ibid., 228.

[272] Ibid., x.

[273] Kane, "Experimental Evidence for More Dimensions."

[274] Glashow, *The Charm of Physics*, 59.

[275] Lodge, *Consciousness and the Novel*.

[276] This quote is from an unattributed review of Emily W. Leider's *Dark Lover: The Life and Death of Rudolph Valentino*, appearing in the "And Bear in Mind" column of the May 18, 2003, *New York Times Book Review*.

[277] Park, "Lacquer on Everything." Park's comments are in review of Mary Yukari Waters' *The Laws of Evening*.

[278] The advertisement "Maharishi's Proposal for Permanent World Peace" is published occasionally, for example: *New York Times*, September 23, 2001, page A29.

[279] Zlotowitz and Scherman, *Bereshis/Genesis*, 29.

[280] Ibid., 36-37.

[281] Ibid., 36.

[282] The 1936 publication in Russian of Born's paper is noted in Chechelnitsky, "Mystery of the 'Magic Number' 137."

[283] The introductory paragraphs of this chapter draw from Gribbin, *Q Is for Quantum*, 52-55, 118, 139, and 141-44.

[284] Casey, "Prime Number Theory and the Fine-Structure Constant."

[285] Anastassov, "Determination of the Fine Structure Constant by ."

[286] Castro, "Fractal strings as an alternative justification for El Naschie's cantorian spacetime and the fine structure constant."

[287] Castro, "Fractal strings as an alternative justification for El Naschie's cantorian spacetime and the fine structure constant" and "A note on transfinite M theory and the fine structure constant."

[288] Olive and Pospelov, "Evolution of the Fine Structure Constant."

[289] Li and Gott, "Inflation in Kaluza-Klein Theory."

[290] de Aquino, "Superparticles from the Initial Universe."

[291] Gupta, Dieckmann, Hadzibabic, and Pritchard, "Contrast Interferometry Using Bose-Einstein Condensates."

[292] This phenomenon has been widely discussed, for example in Barrow and Mota, "Qualitative Analysis of Universes with Varying Alpha"; Battye, Crittenden, and Weller, "Cosmic concordance and the fine structure constant"; Rañada, "On the cosmological variation of the fine structure constant"; Sidharth, "The Holistic Universe and the Variation of the Fine Structure Constant" and "Time-Varying Fine-Structre Constant and Quantum SuperStrings"; and Youm, "A Varying-alpha Brane World Cosmology."

[293] Ter-Kazarian, "Introduction to the Theory of Goyaks."

[294] Tegmark and Vilenkin, "Anthropic predictions for neutrino masses."

[295] Tkachenko, "Electrostatic effects in DNA stretching." Endnote references for examples of possible new physics signals and standard model violations are provided only for phenomena of particularly narrow physicists' focus. Most of these phenomena are being broadly investigated by many physicists throughout the world.

[296] Grossman, Kagan, and Neubert, "Trojan Penguins and Isospin Violation in Hadronic B Decays."

[297] Stecker, "Implications of Ultrahigh Energy Air Showers for Physics and Astrophysics."

[298] Eccles, *Evolution of the Brain*, 190.

[299] The paragraph on the work of the Bogdanovs is based on Armstrong, "The Bogdanov Brothers and their Hoax"; Baez, "The Bogdanov Affair"; Jadczyk and Knight-Jadczyk, "The Bogdanov Singularity"; Johnson, "In Theory, It's True (Or Not)"; Orlowski, "Physics hoaxers discover Quantum Bogosity"; and Overbye, "Are They a) Geniuses or b) Jokers?"

[300] Bearden and Rosenthal, in "On a Testable Unification of Electromagnetics, General Relativity, and Quantum Mechanics," cite the 1903-4 work of physicist E. T. Whittaker.

[301] Jibu and Yasue, "What Is Mind?"

[302] Marcer, "Getting Quantum Theory Off the Rocks."

[303] Lopez de Lema, "Two-Behavior Physics."

[304] Ehresmann and Vanbremeersch, "Emergence Processes Up to Consciousness."

[305] Lee, Jayatilaka, and Kwok, "Virtual Organization."

[306] Bob and Faber, "Quantum Information in Brain Neural Networks and Electroencephalogram."

[307] Laurikainen, "Atoms and Consciousness as Complementary Elements of Reality."

[308] Zaman, "Nature's Psychogenic Forces."

[309] Petrov, ""Psychomas as Mathematical and Topological Images for Elements of Consciousness."

[310] Yasue, "Quantum Monadology."

[311] http://www.consciousness.arizona.edu/quantum-mind.html.

[312] Faisal, quoting Niels Bohr, in "Quantum Chaos, Algorithmic Paradigm and Irreversibility," 47.

[313] Carlip, "Quantum gravity: a progress report," 885-942. The quotes are from page 888.

[314] Smolin, *Three Roads to Quantum Gravity*, 210.

[315] Ibid.

[316] Ibid, 218.

[317] Ibid.

[318] Ibid, 210.

[319] Ibid, 211.

[320] Alonso, Muga, and Sala Mayato, "Comment on 'Foundations of quantum mechanics.'"

[321] Avery, "The Theory of Dimensional Correspondence," 137.

[322] Ibid., 133.

[323] Ibid.

[324] Ibid, 134.

[325] Botta Cantcheff, "Unified field theory from one-particle physics," 2.

[326] This chapter's summary of Botta Cantcheff's work is based on three works of Botta Cantcheff: "Generalized geometries and kinematics for Quantum Gravity," "The 'phenomenological' space time and quantization," and "Unified field theory from one-particle physics."

[327] Spaans, "Topological Dynamics," "Topological Formulation," and "Towards a Topological Formulation."

[328] Spaans, "Observational Tests" and "Topological Extension."

[329] The discussion of Topological Dynamics from here through the end of this chapter is based on Spaans, "On the Topological Nature."

[330] Spaans, "On the Topological Nature," 3-4.

[331] Quotations in the "Implications" section are from Spaans, "On the Topological Nature," 17.

[332] Kafatos and Nadeau, *The Conscious Universe*, 158.

[333] Ibid., 160.

[334] Ibid., 158.

[335] Zohar, "Consciousness and Bose-Einstein Condensates."

[336] Wolf, "On the Quantum Mechanics of Dreams and the Emergence of Self-Awareness."

[337] Hameroff and Penrose, "Orchestrated Reduction of Quantum Coherence in Brain Microtubules."

[338] Matzke, "Consciousness: A New Computational Paradigm."

[339] Insinna, "Synchronicity and Emergent Nonlocal Information in Quantum Systems."

[340] Reflection space, as discussed in this chapter, is based on Sirag, "A Mathematical Strategy for a Theory of Consciousness."

[341] From editors' introduction to Part VIII, "Nonlocal Space and Time," in Hameroff, Kaszniak, and Scott, *Toward a Science of Consciousness: The First Tucson Discussions and Debates*, 541-42.

[342] Sidharth, "The Fractal Universe: From the Planck to the Hubble Scale" and "The Fractal Universe."

[343] Sidharth, "Instantaneous Action at a Distance."

[344] Sidharth, "Duality and Cosmology."

[345] Sidharth, "A New Theory of Geomagnetism."

[346] Sidharth, "A Brief Note on Jupiter's Magnetism."

[347] Sidharth, "The Astronomical Link Between India and the Mayans."

[348] Sidharth, "The Origin of Life in the Universe."

[349] Sidharth, "Neutrino Mass and an Ever Expanding Universe" and "Towards the Unification of Fundamental Interactions."

[350] Sidharth, "Quantized Space-Time and Time's Arrow."

[351] Sidharth, "Issues and Ramifications in Quantum Fractal Space Time" and "Quantum Superstrings and Quantized Fractal Space Time."

[352] Altaisky and Sidharth, "*p*-Adic physics below and above Planck scale."

[353] Sidharth, "Fuzzy Non Commutative Space Time."

[354] Sidharth, "The Elusive Monopole" and ""Monopoles and Solitons."

[355] Sidharth, "The Unification of Electromagnetism and Gravity in the Context of Quantized Fractal Space Time," "'Extended' Particles, Non Commutative Geometry and Unification," "Gravitation and Electromagnetism," and "A Reconciliation of Electromagnetism and Gravitation."

[356] Sidharth, "Gravitation and Electromagnetism II" and "Non-Commutative Geometry, Spin and Quarks."

[357] Sidharth, "Quantum Mechanical Black Holes," "Towards the Unification of Fundamental Interactions," and "'Extended' Particles, Non Commutative Geometry and Unification."

[358] Sidharth, "The Nature of Space Time."

[359] Sidharth, "The Holistic Universe and the Variation of the Fine Structure Constant" and "Time-Varying Fine-Structure Constant and Quantum SuperStrings."

360 Sidharth, "Scale Dependent Dimensionality."

361 Sidharth, "The Scaled Universe" and "Scaled Universe II."

362 Sidharth, "The Cosmology of Fluctuations" and "The New Cosmos."

363 Sidharth, "Oscillator Models for the Universe and Dark Energy."

364 Sidharth, "Quantum, Chaos and the Universe" and "The Emergence of the Planck Scale."

365 Samal, "Can Science 'Explain' Consciousness?" The quotes in the three bullet points that follow are from page 1.

366 Samal, "Can Science 'Explain' Consciousness," 2.

367 Samal, "Speculations on a Unified Theory of Matter and Mind."

368 Ibid., 1.

369 Ibid., 10.

370 The discussion of topological geometrodynamics draws from all of the works of physicist Matti Pitkänen that are listed in the bibliography.

371 Kumar, Ivanov, and Stanley, "Information Entropy and Correlations in Prime Numbers."

372 Derbyshire, *Prime Obsession*, 280-95.

373 Pitkänen "*p*-Adic TGD."

374 Castro, "Is quantum space-time infinite dimensional."

375 Ubriaco, "Fermions on the Field of *p*-Adic Numbers."

376 Djordjevic, Dragovich, and Nesic, "*p*-Adic Quantum Cosmology"; Djordjevic, Dragovich, Nesic, and Volovich, "*p*-Adic and Adelic Minisuperspace Quantum Cosmology"; Dragovich and Nesic, "*p*-Adic and Adelic Generalization of Quantum Cosmology."

377 Khrennikov, "Real Non-Archimedean Structure of Spacetime."

378 Dragovich and Nesic, "On *p*-Adic Numbers in Gravity."

379 Vladimirov and Volovich, "On the Nonlinear Dynamical Equation"; Minahan, "Mode Interactions of the Tachyon Condensate"; Ghoshal and Sen, "Tachyon Condensation and Brane Descent Relations"; Brekke and Freund, "*p*-Adic numbers in physics."

380 Stienstra, "Ordinary Calabi-Yau-3 Crystals."

381 Moeller and Schnabl, "Tachyon condensation"; Moeller and Zwiebach, "Dynamics with Infinitely Many Time Derivatives"; Minahan, "Mode Interactions of the Tachyon Condensate."

382 Khrennikov, "*p*-Adic Quantum-Classical Analogue."

383 Avetisov, Bikulov, and Osipov, "*p*-Adic description of characteristic relaxation"; Khrennikov, "*p*-Adic Stochastics."

384 Khrennikov, "Statistical Interpretation of *p*-Adic Quantum Theories."

385 Albeverio, Khrennikov, and Tirozzi, "*p*-Adic Dynamical Systems and Neural Networks."

386 Limongelli and Loidl, "Rational Number Arithmetic by Parallel p-Adic Algorithms."

387 Castro, "Fractal Strings"; Altaisky and Sidharth, "p-Adic physics below and above Planck scales."

388 Altaisky and Sidharth, "p-Adic physics below and above Planck scales."

389 Dragovich, "Non-Archimedean Geometry and Physics on Adelic Spaces."

390 Chistyakov, "Fractal Geometry for Images of Continuous Map."

391 Dubischar, Gundlach, Khrennikov, and Steinkamp, "Attractors of random dynamical systems."

392 Khrennikov, "Classical and quantum mechanics on information spaces."

393 Khrennikov, "Quantum-like formalism for cognitive measurement."

394 Albeverio, Kloeden, and Khrennikov, "Human Memory as a p-Adic Dynamic System."

395 Khrennikov, "Information Interpretation of p-Adic Physics."

396 Pitkänen, "Genes, memes, qualia, and semitrance."

397 Smilga, "Quantum gravity as Escher's dragon."

398 Ross, *Plex1* and *Plex2*.

399 Huning, "Cognition and Neural Network Modelling.".

400 Conrad, "Anti-Entropy and the Origin of Initial Conditions."

401 Jaffe and Fonck, "Energetics of Social Phenomena."

402 Kwon, "The Structural Changes in the Relations of South and North Korea."

403 Pitkänen, "A model for biophotons."

404 Ibid.

405 Pitkänen, "TGD inspired theory of consciousness with applications to biosystems."

406 Pitkänen, "Doomsday diaries," May 20, 2000, entry.

407 Pitkänen, "Topological geometrodynamics, Part I."

408 Drake co-authored his account of the 1961 conference in Drake and Sobel, *Is Anyone Out There?*, 45-64.

409 Drake and Sobel, *Is Anyone Out There?*, 62.

410 Ulmschneider, *Intelligent Life in the Universe*, 166.

411 Ibid., v and 169-204.

412 The section on Types 1, 2, and 3 worlds draws from Drake and Sobel, *Is Anyone Out There?*, 250, and from Webb, *If the Universe Is Teeming with Aliens . . . Where Is Everybody?*, 2 and 69.

413 Drake and Sobel, *Is Anyone Out There?*, 84-85.

414 Webb, *If the Universe Is Teeming with Aliens . . . Where Is Everybody?*, 51-55.